THE INTERNATIONAL
ENCYCLOPEDIA
OF PHYSICAL CHEMISTRY
AND CHEMICAL PHYSICS

Topic 4.  ELECTRONIC STRUCTURE OF MOLECULES

EDITOR: J. W. LINNETT

Volume 3

THE ELECTRONIC STRUCTURE OF MOLECULES:
THEORY AND APPLICATION TO INORGANIC
MOLECULES

BY

G. DOGGETT

# THE INTERNATIONAL ENCYCLOPEDIA
## OF PHYSICAL CHEMISTRY AND CHEMICAL PHYSICS

# THE INTERNATIONAL ENCYCLOPEDIA
## OF PHYSICAL CHEMISTRY AND CHEMICAL PHYSICS

*Editors-in-Chief*

D. D. ELEY

NOTTINGHAM

F. C. TOMPKINS

LONDON

*List of Topics and Editors*

1. Mathematical Techniques — H. JONES, *London*
2. Classical and Quantum Mechanics — R. McWEENY, *Sheffield*
3. Electronic Structure of Atoms — C. A. HUTCHISON, JR., *Chicago*
4. Electronic Structure of Molecules — J. W. LINNETT, *Cambridge*
5. Molecular Structure and Spectra — Editor to be appointed
6. Kinetic Theory of Gases — E. A. GUGGENHEIM, *Reading* (Deceased)
7. Classical Thermodynamics — D. H. EVERETT, *Bristol*
8. Statistical Mechanics — J. E. MAYER, *La Jolla*
9. Transport Phenomena — J. C. McCOUBREY, *Birmingham*
10. The Fluid State — J. S. ROWLINSON, *London*
11. The Ideal Crystalline State — M. BLACKMAN, *London*
12. Imperfections in Solids — J. M. THOMAS *Aberystwyth*
13. Mixtures, Solutions, Chemical and Phase Equilibria — M. L. McGLASHAN, *Exeter*
14. Properties of Interfaces — D. H. EVERETT, *Bristol*
15. Equilibrium Properties of Electrolyte Solutions — R. A. ROBINSON, *Washington, D.C.*
16. Transport Properties of Electrolytes — R. H. STOKES, *Armidale*
17. Macromolecules — C. E. H. BAWN, *Liverpool*
18. Dielectric and Magnetic Properties — J. W. STOUT, *Chicago*
19. Gas Kinetics — A. F. TROTMAN-DICKENSON, *Cardiff*
20. Solution Kinetics — R. M. NOYES, *Eugene*
21. Solid and Surface Kinetics — F. C. TOMPKINS, *London*
22. Radiation Chemistry — Editor to be appointed

# THE ELECTRONIC STRUCTURE OF MOLECULES: THEORY AND APPLICATION TO INORGANIC MOLECULES

BY

## G. DOGGETT

UNIVERSITY OF YORK

PERGAMON PRESS

OXFORD · NEW YORK · TORONTO
SYDNEY · BRAUNSCHWEIG

Pergamon Press Ltd., Headington Hill Hall, Oxford
Pergamon Press Inc., Maxwell House, Fairview Park, Elmsford,
New York 10523
Pergamon of Canada Ltd., 207 Queen's Quay West, Toronto 1
Pergamon Press (Aust.) Pty. Ltd., 19a Boundary Street,
Rushcutters Bay, N.S.W. 2011, Australia
Vieweg & Sohn GmbH, Burgplatz 1, Braunschweig

First edition 1972

Library of Congress Catalog Card No. 75–171466

*Printed in Great Britain by*
Adlard & Son Ltd., Bartholomew Press, Dorking
08 016588 5

# INTRODUCTION

THE International Encyclopedia of Physical Chemistry and Chemical Physics is a comprehensive and modern account of all aspects of the domain of science between chemistry and physics, and is written primarily for the graduate and research worker. The Editors-in-Chief, Professor D. D. ELEY, Professor J. E. MAYER and Professor F. C. TOMPKINS, have grouped the subject matter in some twenty groups (General Topics), each having its own editor. The complete work consists of about one hundred volumes, each volume being restricted to around two hundred pages and having a large measure of independence. Particular importance has been given to the exposition of the fundamental bases of each topic and to the development of the theoretical aspects; experimental details of an essentially practical nature are not emphasized although the theoretical background of techniques and procedures is fully developed.

The Encyclopedia is written throughout in English and the recommendations of the International Union of Pure and Applied Chemistry on notation and cognate matters in physical chemistry are adopted. Abbreviations for names of journals are in accordance with *The World List of Scientific Periodicals*.

# CONTENTS

|  |  | PAGE |
|---|---|---|
| PREFACE | | xi |
| UNITS AND NOTATION | | xiii |
| CHAPTER 1 | ORBITAL THEORIES OF ELECTRONIC STRUCTURE | 1 |
| 1.1 | Introduction | 1 |
| 1.2 | The separation of electronic and nuclear motions | 2 |
| 1.3 | Determination of approximate electronic wave functions | 10 |
| 1.4 | The valence-bond method | 11 |
| 1.5 | The method of molecular orbitals | 36 |
| 1.6 | A brief résumé of open-shell molecular orbital theory | 47 |
| 1.7 | Semi-empirical molecular orbital theory | 49 |
| 1.8 | The electron density function | 52 |
| CHAPTER 2 | THE ELECTRONIC STRUCTURE OF SOME MOLECULES CONTAINING A CENTRAL SECOND-ROW ATOM | 67 |
| 2.1 | Introduction | 67 |
| 2.2 | Sulphur hexafluoride, $SF_6$ | 73 |
| 2.3 | Sulphur tetrafluoride, $SF_4$ | 88 |
| 2.4 | Sulphur pentafluoride, $SF_5$ | 91 |
| 2.5 | Phosphorus pentafluoride, $PF_5$ | 92 |
| 2.6 | Chlorine trifluoride, $ClF_3$ | 92 |
| CHAPTER 3 | THE ELECTRONIC STRUCTURE OF TRANSITION METAL ION COMPLEXES | 98 |
| 3.1 | Introduction | 98 |
| 3.2 | The crystal field model: $d^1$ complexes | 103 |
| 3.3 | The treatment of vibronic interactions | 109 |
| 3.4 | $d^2$ octahedral complexes | 114 |
| 3.5 | An appraisal of crystal field theory | 122 |
| 3.6 | Molecular orbital theory of complex ions | 126 |
| 3.7 | The problem of complex ions with open-shell electronic structures | 132 |
| CHAPTER 4 | THE ELECTRONIC STRUCTURE OF XENON FLUORIDES | 136 |
| 4.1 | Introduction | 136 |
| 4.2 | The valence bond model | 137 |
| | (a) Xenon difluoride, $XeF_2$ | 137 |
| | (b) Xenon tetrafluoride, $XeF_4$ | 141 |
| | (c) Xenon hexafluoride, $XeF_6$ | 144 |

PAGE

4.3           The molecular orbital model       149
                 (a) Xenon difluoride, $XeF_2$       149
                 (b) Xenon tetrafluoride, $XeF_4$       150
                 (c) Xenon hexafluoride, $XeF_6$       152

APPENDIX I     CALCULATION OF THE DIAGONAL MATRIX ELEMENT OF THE MOLECULAR ELECTRONIC HAMILTONIAN FOR A WAVE FUNCTION GIVEN AS AN ANTISYMMETRIZED PRODUCT OF ONE-ELECTRON FUNCTIONS (NOT NECESSARILY ORTHOGONAL)       157

APPENDIX II     CHARACTER TABLES FOR $O_h$ AND $T_d$ POINT GROUPS       162

APPENDIX III     SYMMETRY ADAPTED COMBINATIONS OF ATOMIC ORBITALS FOR AN OCTAHEDRAL MOLECULE (POINT GROUP $O_h$)       163

AUTHOR INDEX       165

SUBJECT INDEX       169

# PREFACE

THE aim of this book is to provide Chemistry students in their final year, or first year of postgraduate study, with a reasonably detailed account of the methods used for investigating the electronic structure of inorganic molecules and ions. The exposition has been kept as straightforward as possible so that the student can clearly discern the principles underlying the various model calculations. This approach has necessitated a lengthy introductory chapter on basic methods; but without this it is difficult for a beginner to develop an objective view of the problems inherent in a molecular structure calculation.

It is unfortunate, perhaps, that, in the space available, it has not been possible to say anything, for example, about the magnetic properties of transition metal ion complexes, or relevant aspects of the solid state—just to mention two omissions. Deficiencies such as these are unavoidable, unless the text is expanded considerably. In any case, my main concern has been to focus attention on the basic problems of determining the electronic structure of isolated molecules: the extra problems involved when the molecules are subjected to additional perturbations arising from external electric or magnetic fields are immense, and do not really fall within the confines of this section of the Encyclopedia.

This book was written while I was on the staff of the Chemistry Department, University of Glasgow, and I am indebted to my former colleagues, Dr. B. C. Webster and Dr. R. V. Emanuel, and research students for many stimulating discussions on various aspects of the manuscript. I would also like to thank Dr. T. Thirunamachandran for a critical reading of the manuscript, and my wife for her forbearance during the whole course of the writing and typing of the manuscript.

*Glasgow, May* 1970 G. DOGGETT

xi

# UNITS AND NOTATION

*Lengths* are given in nanometres (nm):

$$1 \text{ nm} = 10^{-9} \text{ m} = 10 \text{ Å}.$$

$$1 \text{ a.u.} = a_0 = \frac{\epsilon_0 h^2}{\pi m e^2} = 0 \cdot 05292 \text{ nm} = 0 \cdot 5292 \text{ Å}.$$

($\epsilon_0 = 8 \cdot 854 \times 10^{-12} \text{ Fm}^{-1}$, the permittivity of free space.)

*Energies* are given in either atomic units (a.u.) or attojoules (aJ):

$$1 \text{ a.u.} = \frac{e^2}{4\pi\epsilon_0 a_0} = 4 \cdot 359 \text{ aJ} = 27 \cdot 21 \text{ eV}.$$

$$1 \text{ aJ} = 10^{-18} \text{ J} = 6 \cdot 242 \text{ eV} = 5 \cdot 035 \times 10^4 \text{ cm}^{-1}.$$

*Dipole moments* are given in Debyes:

$$1 \text{ D} = 10^{-18} \text{ e.s.u. cm} = 0 \cdot 3336 \times 10^{-29} \text{ Cm}.$$

All operators are represented by upper- or lower-case italic letters with a circumflex accent: for example, $\hat{H}$ or $\hat{s}_z(m)$.

$$\sum'_{i,j} A_{ij}$$

indicates summation over $i$ and $j$, excluding the terms with $i = j$;

$$\sum'_{j} A_{ji}$$

indicates summation over $j$, excluding the term with $j = i$.

A superscript * indicates the complex conjugate.

Matrices are denoted by bold roman letters: for example, $\mathbf{S}$. The $ij$th element of $\mathbf{S}$ is denoted by either $(\mathbf{S})_{ij}$ or $S_{ij}$. The transpose of $\mathbf{S}$ is denoted by $\mathbf{S}'$, where $(\mathbf{S}')_{ij} = (\mathbf{S})_{ji}$. The adjoint of $\mathbf{S}$ is denoted by $\mathbf{S}^+$, where $(\mathbf{S}^+)_{ij} = (\mathbf{S}^*)_{ji}$. The trace of $\mathbf{S}$ is denoted by tr $\mathbf{S}$, and is given by the sum of the diagonal elements of $\mathbf{S}$ ($\mathbf{S}$ assumed square).

$\hat{P}$ is an operator which permutes electron labels.

The collection of symmetry operations associated with a general molecular point group is denoted by $G$.

$D_\Gamma(\hat{R})_{ji}$ is the $ji$th element of the matrix representing the symmetry operation $\hat{R}$ in the irreducible representation $\Gamma$.

$\chi_\Gamma(\hat{R})$ is the character associated with the symmetry operation $\hat{R}$ in the irreducible representation $\Gamma$.

The following convention is adopted for displaying atomic orbitals in the figures representing electron-pairing schemes for molecular electronic structure:

> $s$ atomic orbitals are indicated by a single circular lobe;
>
> $p$ atomic orbitals are indicated by two (non-touching) circular lobes;
>
> $d$ atomic orbitals are indicated by four (non-touching) egg-shaped lobes ($t_{2g}$ variety);
>
> hybrid atomic orbitals are indicated by pear-shaped lobes.

This convention is necessary only for displaying the types of atomic orbital involved in the bonding, and the size of the lobes is not indicative of the spatial extension of the atomic orbitals.

# CHAPTER 1

# ORBITAL THEORIES OF ELECTRONIC STRUCTURE

## 1.1. Introduction

A detailed understanding of the electronic structure of many-electron atoms and molecules still remains one of the basic problems in the application of wave mechanics to systems of chemical interest: this situation arises because the Schrödinger equation can be solved exactly only for one-electron atoms. Thus, in general, it is necessary to find approximate wave functions which simulate the exact wave function as well as possible. As described in Vol. 1 this usually means incorporating linear and non-linear parameters into the approximate wave function, and then applying the variation theorem to determine their optimum values. This technique has been used with great success in the study of some two-electron systems.

Pekeris[1] has calculated the binding energy of He more accurately than the experimentally determined value; while Kołos and Wolniewicz[2] have calculated the dissociation energy of $H_2$ to within the limits of error of the experimental value (see Herzberg, Bibliography). The numbers of variationally determined parameters in these calculations were 1078 and 55, respectively; a factor which precludes any generalization of these particular methods to molecules of chemical interest. Thus, rather than trying to calculate the binding energy of a transition metal ion complex which, in any case, may be smaller than the errors inherent in the calculation, it is more profitable to calculate changes in selected molecular properties within a series of closely related molecules. Any errors arising from the basic assumptions, or approximations, should then remain constant, so allowing the trend of a particular property to attain some meaning. An example of this approach is in the study of a particular band, or bands, in the electronic absorption spectrum of an octahedrally coordinated transition metal ion for different choices of ligand.

For very large molecules, the possibility of performing non-empirical wave mechanical calculations is quite remote, and further approximations are inevitably invoked. For example, it is usual to separate the electrons into groups: the valence electrons are considered in one group, while all the remaining electrons provide a non-polarizable core—just as $\sigma$-electrons do in $\pi$-electron theories of conjugated hydrocarbons. This separation is often necessary for making calculations tractable, and it is not a consequence of a particular symmetry requirement. In most cases, though, it represents a reasonable approximation as many properties of chemical interest reflect changes in the distribution of the valence electrons. However, the core–valence electron separation requires handling carefully: particularly when the molecular wave function explicitly involves the valence electrons alone. Even though most experimental situations are dominated by changes within the group of valence electrons, there are some interesting cases where the effects of valence and core electrons cannot be separated completely: as found, for example, in the Mössbauer effect, X-ray emission spectra, contact spin coupling, and in understanding the finer details of nuclear quadrupole resonance spectroscopy.

Further simplifications can often be made for molecules possessing a high degree of symmetry. For example, an extensive understanding of the chemistry of planar unsaturated hydrocarbons has been achieved through the assumption of $\sigma$–$\pi$ separability. Although the validity of this basic assumption is now in doubt, its use has undeniably correlated a large amount of experimental data (see Vol. 2).

As far as non-linear and non-planar inorganic molecules are concerned, and these include the highly symmetrical octahedral $XY_6$ and tetrahedral $XY_4$ molecules, it is not formally possible to effect a $\sigma$–$\pi$ separation: nevertheless, it is often very useful to assume that the electronic structure of these symmetrical molecules can be discussed in terms of $\sigma$ and $\pi$ contributions to the overall bonding. In this context, of course, it is strictly only meaningful to use the terms $\sigma$ and $\pi$ when discussing the ligand–central atom interactions within a chosen bond: for it is only then that the cylindrical symmetry permits orbitals to be classified as either $\sigma$ or $\pi$ type.

It has so far been tacitly assumed that the properties of a particular molecule are adequately described in terms of the various electrons moving with respect to a fixed nuclear framework; that is, the electronic and nuclear motions have been separated through use of the Born–Oppenheimer[3] approximation, a preliminary discussion of which has already been given in Vol. 1. The Born–Oppenheimer approximation may not always be applicable and, when this is so, the consequences are of considerable interest. For this reason it is important to examine the assumptions implicit in the approximation, and to develop the necessary analysis in some detail.

## 1.2.   The separation of electronic and nuclear motions

In Vol. 1 it was shown how the use of the Born–Oppenheimer approximation enables each bound state of a molecule to be represented by a product wave function of the form

$$\Phi_m^k = \tilde{\Psi}_m(\mathbf{r}\,;\,\mathbf{R},\,\mathbf{s})\Lambda_m^k(\mathbf{R}). \tag{1.1}$$

**r**, **R**, **s** represent the complete sets of electron, nuclear and electron spin coordinates, respectively; and $m$, $k$ label different electronic and nuclear wave functions, respectively. Thus for each electronic state $\tilde{\Psi}_m$ there is a set of nuclear wave functions $\Lambda_m^k$ ($k = 1, 2 \ldots$) describing the vibrational and rotational states of the molecule.

The total spinless molecular Hamiltonian, $\mathscr{H}$, is given by

$$\mathscr{H} = -\frac{1}{2}\sum_i \nabla_i^2 - \sum_{i,\alpha} \frac{Z_\alpha}{r_{\alpha i}} + \sum_{i<j} \frac{1}{r_{ij}} + \sum_{\alpha<\beta} \frac{Z_\alpha Z_\beta}{R_{\alpha\beta}} - \frac{1}{2}\sum_\alpha \frac{m}{M_\alpha} \nabla_\alpha^2 \quad (1.2)$$

where $i, j$ label electrons and $\alpha, \beta$ label nuclei. The form of (1.2) implies energy is measured in units of $e^2/(4\pi\epsilon_0 a_0)$ (Hartrees), where $a_0$ is the atomic unit of length ($a_0 = \epsilon_0 h^2/\pi m e^2 = 1$ Bohr).

$\tilde{\Psi}_m$ is assumed to be a solution of the electronic Schrödinger equation

$$\hat{H}\tilde{\Psi}_m \equiv \left(-\frac{1}{2}\sum_i \nabla_i^2 - \sum_{i,\alpha} \frac{Z_\alpha}{r_{\alpha i}} + \sum_{i<j} \frac{1}{r_{ij}}\right) \tilde{\Psi}_m = E_m(\mathbf{R})\tilde{\Psi}_m \quad (1.3)$$

in which the nuclei are held in the fixed configuration **R**. The nuclear coordinates therefore appear as parameters in the electronic wave function, merely defining the nuclear configuration. For present purposes, $E_m(\mathbf{R})$ is assumed to be non-degenerate; thereby enabling $\tilde{\Psi}_m$ to be chosen real without any loss of generality.

The equation determining $\Lambda_m^k$ is found by straightforward application of the variation theorem using (1.1) and (1.2):

$$\delta \iint \Phi_m^k(\mathscr{H} - \varepsilon_m^k)\Phi_m^k \, d\tau \, d\mathbf{R} = 0, \quad (1.4)$$

where $d\tau = d\mathbf{r} \, d\mathbf{s} = d\mathbf{r}_1 \, ds_1 \, d\mathbf{r}_2 \, ds_2 \ldots = dx_1 \, dy_1 \, dz_1 \, ds_1 \, dx_2 \ldots$. Substitution of (1.1) and (1.2) in (1.4) then gives

$$\iint \tilde{\Psi}_m \delta\Lambda_m^k \left[\hat{H} + \frac{1}{2}\sum_\alpha \left(-\frac{m}{M_\alpha}\right)\nabla_\alpha^2 + \sum_{\alpha<\beta} \frac{Z_\alpha Z_\beta}{R_{\alpha\beta}} - \varepsilon_m^k\right] \tilde{\Psi}_m \Lambda_m^k \, d\tau \, d\mathbf{R} = 0$$

which, after using (1.3) and integrating over the electronic and spin coordinates, becomes

$$\int \delta\Lambda_m^k \left[E_m(\mathbf{R}) + \sum_{\alpha<\beta} \frac{Z_\alpha Z_\beta}{R_{\alpha\beta}} + \frac{1}{2}\sum_\alpha \left(-\frac{m}{M_\alpha}\right) \int \tilde{\Psi}_m \nabla_\alpha^2 \tilde{\Psi}_m \, d\tau \right.$$
$$\left. + \frac{1}{2}\sum_\alpha \left(-\frac{m}{M_\alpha}\right)\nabla_\alpha^2 - \varepsilon_m^k\right] \Lambda_m^k \, d\mathbf{R} = 0, \quad (1.5)$$

where the term involving the integral $\int \tilde{\Psi}_m \nabla_\alpha \tilde{\Psi}_m \, d\tau$ vanishes since

$$\nabla_\alpha \int \tilde{\Psi}_m \tilde{\Psi}_m \, d\tau = 2 \int (\nabla_\alpha \tilde{\Psi}_m)\tilde{\Psi}_m \, d\tau = 0.$$

Now $\delta\Lambda_m^k$ represents an arbitrary variation, so the vanishing of (1.5) requires

$$\hat{H}_{\text{nuc}}^m \Lambda_m^k \equiv \left[ E_m(\mathbf{R}) + \sum_{\alpha<\beta} \frac{Z_\alpha Z_\beta}{R_{\alpha\beta}} + \frac{1}{2} \sum_\alpha \left( -\frac{m}{M_\alpha} \right) \int \tilde{\Psi}_m \nabla_\alpha^2 \tilde{\Psi}_m \, d\tau \right.$$
$$\left. + \frac{1}{2} \sum_\alpha \left( -\frac{m}{M_\alpha} \right) \nabla_\alpha^2 \right] \Lambda_m^k = \mathcal{E}_m^k \Lambda_m^k, \quad (1.6)$$

in which the electronic energy, $E_m(\mathbf{R})$, appears as a potential energy term in the effective Hamiltonian for nuclear motion. The total molecular energy is therefore given by

$$\mathcal{E}_m^k = \iint \Phi_m^k \hat{\mathscr{H}} \Phi_m^k \, d\tau \, d\mathbf{R} = \iint \tilde{\Psi}_m \Lambda_m^k \hat{\mathscr{H}} \tilde{\Psi}_m \Lambda_m^k \, d\tau \, d\mathbf{R}$$
$$= \int \Lambda_m^k \hat{H}_{\text{nuc}}^m \Lambda_m^k \, d\mathbf{R}.$$

where $\Lambda_m^k$ satisfies (1.6).

The molecular wave functions (1.1) are not eigenfunctions of $\hat{\mathscr{H}}$: this is most readily seen by operating with $\hat{\mathscr{H}}$ on any one of the wave functions (1.1).

$$\hat{\mathscr{H}}(\tilde{\Psi}_m \Lambda_m^k) = \left( \hat{H} + \frac{1}{2} \sum_\alpha \left( -\frac{m}{M_\alpha} \right) \nabla_\alpha^2 + \sum_{\alpha<\beta} \frac{Z_\alpha Z_\beta}{R_{\alpha\beta}} \right) \tilde{\Psi}_m(\mathbf{r}, \mathbf{R}) \Lambda_m^k(\mathbf{R})$$

$$= \tilde{\Psi}_m \left( E_m(\mathbf{R}) + \sum_{\alpha<\beta} \frac{Z_\alpha Z_\beta}{R_{\alpha\beta}} + \frac{1}{2} \sum_\alpha \left( -\frac{m}{M_\alpha} \right) \nabla_\alpha^2 \right) \Lambda_m^k(\mathbf{R})$$
$$+ \sum_\alpha \left( -\frac{m}{M_\alpha} \right) (\nabla_\alpha \tilde{\Psi}_m) \cdot (\nabla_\alpha \Lambda_m^k) + \frac{1}{2} \sum_\alpha \left( -\frac{m}{M_\alpha} \right) (\nabla_\alpha^2 \tilde{\Psi}_m) \Lambda_m^k$$

$$= \tilde{\Psi}_m \left( E_m(\mathbf{R}) + \sum_{\alpha<\beta} \frac{Z_\alpha Z_\beta}{R_{\alpha\beta}} + \frac{1}{2} \sum_\alpha \left( -\frac{m}{M_\alpha} \right) \right.$$
$$\left. \times \left[ \nabla_\alpha^2 + \int \tilde{\Psi}_m \nabla_\alpha^2 \tilde{\Psi}_m \, d\tau \right] \right) \Lambda_m^k$$
$$+ \sum_\alpha \left( -\frac{m}{M_\alpha} \right) \left[ (\nabla_\alpha \tilde{\Psi}_m) \cdot (\nabla_\alpha \Lambda_m^k) + \frac{1}{2} (\nabla_\alpha^2 \tilde{\Psi}_m) \Lambda_m^k \right.$$
$$\left. - \frac{1}{2} \Lambda_m^k \tilde{\Psi}_m \int \tilde{\Psi}_m \nabla_\alpha \tilde{\Psi}_m \, d\tau \right]$$

$$= \mathcal{E}_m^k \tilde{\Psi}_m \Lambda_m^k + \sum_\alpha \left( -\frac{m}{M_\alpha} \right) \left[ (\nabla_\alpha \tilde{\Psi}_m) \cdot (\nabla_\alpha \Lambda_m^k) \right.$$
$$+ \frac{1}{2} (\nabla_\alpha^2 \tilde{\Psi}_m) \Lambda_m^k$$
$$\left. - \frac{1}{2} \Lambda_m^k \tilde{\Psi}_m \int \tilde{\Psi}_m \nabla_\alpha^2 \tilde{\Psi}_m \, d\tau \right]. \quad (1.7)$$

The terms in square brackets cause the wave functions (1.1) to be coupled together. For example, a typical coupling is represented by the matrix element

$$\iint \Phi_n^k \mathscr{H} \Phi_m^l \, d\tau \, d\mathbf{R}$$

which, after using (1.7), becomes

$$\varepsilon_m^l \cdot \delta_{mn}\delta_{kl} + \int \Lambda_n^k \left[ \int \tilde{\Psi}_n \sum_\alpha \left( -\frac{m}{M_\alpha} \right) (\nabla_\alpha \tilde{\Psi}_m) \, d\tau \right] \cdot (\nabla_\alpha \Lambda_m^l) \, d\mathbf{R}$$

$$+ \frac{1}{2} \int \Lambda_n^k \left[ \int \tilde{\Psi}_n \sum_\alpha \left( -\frac{m}{M_\alpha} \right) (\nabla_\alpha^2 \tilde{\Psi}_m) \, d\tau \right] \Lambda_m^l \, d\mathbf{R}$$

$$- \delta_{mn} \int \Lambda_n^k \left[ \frac{1}{2} \int \tilde{\Psi}_m \sum_\alpha \left( -\frac{m}{M_\alpha} \right) (\nabla_\alpha^2 \tilde{\Psi}_m) \, d\tau \right] \Lambda_m^l \, d\mathbf{R}. \qquad (1.8)$$

Thus there are no matrix elements between different nuclear states associated with the same (non-degenerate) electronic state ($m = n$). But the wave functions are coupled together if $m \neq n$: this also includes the situation in which $\tilde{\Psi}_n$, $\tilde{\Psi}_m$ are degenerate. Now so long as the energies of $\tilde{\Psi}_n$ and $\tilde{\Psi}_m$ are well separated, the effect of the coupling terms will be small as the electronic wave functions vary slowly with changes in the nuclear coordinates. In this situation the coupling terms are usually ignored. The molecule then remains on one electronic potential energy surface, $E_m(\mathbf{R})$, for all motions of the nuclei. This is the basis of the "adiabatic" approximation as proposed by Born and Oppenheimer.[3] However, for degenerate, or near degenerate, electronic states the coupling terms may become very large and invalidate the Born–Oppenheimer approximation. The reason for this behaviour is most readily seen by examining a different form for the integral $\int \tilde{\Psi}_n \nabla_\alpha \tilde{\Psi}_m \, d\tau$, which appears in the matrix element (1.8).

First, the operator $\nabla_\alpha$ is applied to each side of (1.3):

$$(\nabla_\alpha \hat{H}) \tilde{\Psi}_n + \hat{H}(\nabla_\alpha \tilde{\Psi}_n) = [\nabla_\alpha E_n(\mathbf{R})]\tilde{\Psi}_n + E_n(\mathbf{R})(\nabla_\alpha \tilde{\Psi}_n).$$

Multiplication of both sides by $\tilde{\Psi}_m$, followed by integration over space and spin coordinates, then gives

$$\int \tilde{\Psi}_m(\nabla_\alpha \hat{H})\tilde{\Psi}_n \, d\tau + \int \tilde{\Psi}_m \hat{H}[\nabla_\alpha \tilde{\Psi}_n] \, d\tau = E_n \int \tilde{\Psi}_m \nabla_\alpha \tilde{\Psi}_n \, d\tau$$

which, from the Hermitian nature of $\hat{H}$, becomes

$$(E_n - E_m)^{-1} \int \tilde{\Psi}_m(\nabla_\alpha \hat{H})\tilde{\Psi}_n \, d\tau = \int \tilde{\Psi}_m \nabla_\alpha \tilde{\Psi}_n \, d\tau. \qquad (1.9)$$

Equation (1.9) shows clearly why the coupling terms cannot be ignored when the electronic states are not well separated in energy: for then, the electronic wave functions are very sensitive to changes in the nuclear coordinates.

The coupling between the various molecular wave functions, (1.1), has the effect of contaminating a given non-degenerate wave function, for example

$\Phi_m^k$, with small amounts of the higher energy states $\Phi_n^l$ ($n \neq m$). The improved wave function is therefore represented by

$$\Phi' = C_{mk}(\mathbf{R})\widetilde{\Psi}_m^k \Lambda_m^k + \sum_{l,\, n\,(n \neq m)} C_{nl}(\mathbf{R})\widetilde{\Psi}_n^l \Lambda_n^l \tag{1.10}$$

where the value of $C_{ps}(\mathbf{R})$ gives the degree of mixing of $\Phi_p^s$ with the reference state $\Phi_m^k$. Since the coefficients $C_{ps}(\mathbf{R})$ depend only upon the nuclear co-ordinates $\mathbf{R}$, they are often absorbed into the $\Lambda_p^s(\mathbf{R})$:

$$\Phi' = \widetilde{\Psi}_m^k \Xi_m^k + \sum_{l,\, n\,(n \neq m)} \widetilde{\Psi}_n^l \Xi_n^l$$

and the expansion coefficients, $\Xi_p^s$, are then optimized by application of the variation theorem. As described in Vol. 1, the improved state energies are obtained from the solution of a secular determinant, with elements given by (1.8). The secular equations are then solved for the optimum values of the $\Xi_p^s(\mathbf{R})$.

If the coupling terms are small, but non-negligible, the improved energies may be found by applying perturbation theory. But for cases of degeneracy, or near degeneracy, it is necessary to partially diagonalize the energy matrix (and correspondingly the secular determinant) to remove large off-diagonal terms before applying perturbation theory. When there is degeneracy present, particularly in the ground state, the coupling between the nuclear and electronic motions causes the molecule to distort, thereby relieving the degeneracy: this phenomenon is referred to as the Jahn–Teller effect for non-linear molecules, and the Renner effect for linear polyatomic molecules. The distortion may persist if the splitting of the degenerate levels is greater than the zero point energy for vibrational motion; in which case the molecule vibrates about a new equilibrium configuration. This is in essence the static Jahn–Teller effect. The dynamic Jahn–Teller effect is more involved as the molecule is continuously tunnelling between a number of equivalent configurations possessing the same energy. The reader is referred to the standard references for a more complete discussion of this problem.[4-6]

The assumption of the Born–Oppenheimer approximation leads to a situation in which the molecule is characterized by a set of vibrational and rotational states associated with each electronic state (see Vol. 1). Experimental evidence indicates that in general this is a good approximation. However, for many-electron molecules the solution of (1.3) is possible only when some form of approximate electronic wave function is used. In practice, the $\widetilde{\Psi}_m$ are frequently constructed out of limited sets of atomic orbitals centred on each nucleus in the molecule. These approximate wave functions usually contain a number of linear and non-linear parameters which can then be varied to minimize the electronic energy, and hence approach the exact energy as closely as possible. Approximations to $\widetilde{\Psi}_m$ will be denoted by $\Psi_m$.

The use of approximate electronic wave functions, which are of course not eigenfunctions of $\hat{H}$, results in similar complications to those experienced when working with the approximate molecular wave functions (1.1). Although the wave functions $\Psi_m$ do not diagonalize $\hat{H}$, a model Hamiltonian, $\hat{H}_{eff}$, can usually be found which is diagonal with respect to the set of expansion functions, $\Psi_m$: thus, on writing the operator identity

$$\hat{H} = \hat{H} - \hat{H}_{eff} + \hat{H}_{eff} = \hat{H}_{eff} + \hat{V}$$

it is readily seen that the coupling between the various approximate electronic states arises through $\hat{V}$, since only $\hat{H}_{eff}$ is diagonal with respect to the set of functions $\Psi_m$:

$$\int \Psi_m \hat{H}_{eff} \Psi_n \, d\tau = \delta_{mn} E_m,$$

$$\int \Psi_m \hat{V} \Psi_n \, d\tau = V_{nm},$$

where $V_{nm}$ is in general not equal to zero. The form of $\hat{H}_{eff}$ depends upon the nature of the approximations used for obtaining $\Psi_m$. However, irrespective of the choice of $\Psi_m$, an improved set of wave functions can always be found by allowing for the coupling between the approximate wave functions, induced by the non-coincidence of $\hat{H}$ and $\hat{H}_{eff}$:

$$\Psi'_m = \sum_p a_{pm} \Psi_p. \tag{1.11}$$

The coefficients, $a_{pm}$, and the improved energies, are found in the usual way by application of the variation theorem (see Vol. 1). Quite frequently a number of the matrix elements in the secular determinant are zero on account of symmetry, or for other reasons emanating from the chosen form of $\hat{H}_{eff}$.

The method described above for handling approximate wave functions involves searching for a convenient partitioning of the total Hamiltonian, $\hat{H}$, in the form

$$\hat{H} = \sum_i \hat{H}_i.$$

The complete solution of the problem is then approached in a step-wise manner by first diagonalizing $\hat{H}_1 + \hat{H}_2$ with the wave functions which diagonalize $\hat{H}_1$; the resulting wave functions are then used to diagonalize $\hat{H}_1 + \hat{H}_2 + \hat{H}_3$, and so on. This particular technique is so widely used that it is worth considering a further example which illustrates some different aspects of the partitioning procedure: this is the problem of treating the effects of spin–orbit coupling in atoms. The example is chosen because it illustrates a technique which will be used in discussing aspects of crystal field theory in Chapter 3.

The Hamiltonian for the many-electron atom includes the kinetic and potential energy operators for all the electrons, as well as additional relativistic terms which are usually ignored in the case of light atoms. The present discussion is concerned only with the relativistic term corresponding to the effects of spin–orbit coupling. The total Hamiltonian is therefore partitioned in the following way: $\qquad \hat{H} = \hat{H}_1 + \hat{H}_2 + \hat{H}_3,$

where $\hat{H}_1$ is the sum of one-electron operators corresponding to kinetic and electron-nuclear potential energies; $\hat{H}_2$ describes the interelectronic repulsions, and $\hat{H}_3$ describes the effects of spin–orbit coupling.†

In many instances, where the effects produced by $\hat{H}_3$ are small, the set of approximate wave functions $\Psi_m$, which diagonalize $\hat{H}_1$, can be used to diagonalize $\hat{H}_1 + \hat{H}_2$: this yields an improved set of electronic wave functions $\Psi'_m$ (see (1.11)). The effects of $\hat{H}_3$ are then determined by expanding the wave functions for the complete problem in terms of the $\Psi'_m$. Off-diagonal elements in the secular determinant now involve $\hat{H}_3$ only, on account of the particular choice of expansion functions:

$$\int \Psi'_m \hat{H} \Psi'_n \, d\tau = \int \Psi'_m \hat{H}_3 \Psi'_n \, d\tau.$$

The solution of the new secular equations then proceeds in the usual way to yield the improved energies and wave functions. It can be argued that this is a very arduous procedure, because the diagonalization of $\hat{H}$, with respect to the original set of functions $\Psi_m$, requires the solution of only one set of secular equations. However, it is often more profitable to diagonalize $\hat{H}$ in stages, as the effects of adding successive terms to the Hamiltonian are then more clearly discernible.

In situations where the effects of $\hat{H}_2$ and $\hat{H}_3$ may be comparable, it is no longer obvious whether $\hat{H}_1 + \hat{H}_2$ or $\hat{H}_1 + \hat{H}_3$ should be diagonalized first; and then followed by the diagonalization of $\hat{H}_3$ or $\hat{H}_2$, respectively. However, the same sequence of energy levels is obtained in both cases, although the wave functions may look different. A simple example will make this point clear.

In the case of intermediate coupling, as required in the treatment of moderately heavy atoms, the diagonalization of $\hat{H}_1 + \hat{H}_2$ gives rise to the Russell–Saunders term wave functions, $\Psi'_m$ (spin–orbit coupling neglected), which are characterized by the quantum numbers $L$, $S$, $M_S$, $M_L$, $(M_J)$. These functions are then used to diagonalize the spin–orbit part of the Hamiltonian, represented by $\hat{H}_3$. $L$ and $S$ (and $M_L$, $M_S$) are no longer good quantum numbers, and the resulting solutions have $J$, $M_J$ as good quantum numbers only. Now if $\hat{H}_1 + \hat{H}_3$ is diagonalized first, yielding the set of wave functions $\Psi''_m$, this corresponds to the $j$–$j$ coupling limit in which the effects of spin-orbit coupling dominate the electrostatic interactions, represented by $\hat{H}_2$. The diagonalization of $\hat{H}_2$ then produces states again having $J$, $M_J$ as good quantum numbers. Thus, the wave functions which diagonalize $\hat{H}$ can be expressed in terms of either the $j$–$j$ or the $L$–$S$ coupling schemes. And since

† $\hat{H}$ is often partitioned in the form $(\hat{H}_1 + \hat{V}_{\text{eff}}) + (\hat{H}_2 - \hat{V}_{\text{eff}}) + \hat{H}_3$, where $\hat{V}_{\text{eff}}$ represents the sum of the average potential energies of interaction of each electron with the remaining electrons; thereby allowing for nuclear screening effects in determining the $\Psi_m$.

both sets of wave functions give rise to the same sequence of energy levels, the two sets of wave functions must be related by means of a unitary transformation: a result which is now derived.

Let $\Psi_m'''$ denote the wave function corresponding to the electronic state with defined values of $J$, $M_J$. The energy of this state is given by

$$E = \frac{\int \Psi_m'''{}^* \hat{H} \Psi_m''' \, d\tau}{\int \Psi_m'''{}^* \Psi_m''' \, d\tau} = \frac{\sum\limits_{p,\,q} c_{pm}^* c_{qm} \int \Psi_p'{}^* \hat{H} \Psi_q' \, d\tau}{\sum\limits_{p,\,q} c_{pm}^* c_{qm} \int \Psi_p'{}^* \Psi_q' \, d\tau}$$

$$= \sum\limits_{p,\,q} c_{pm}^* H_{pq} c_{qm} \Big/ \sum\limits_{p,\,q} c_{pm}^* \delta_{qp} c_{qm}$$

where the $\Psi_m'$ are assumed to be orthonormal. The numbers $c_{pm}$ and $H_{pq}$ can be regarded as elements of the matrices denoted by $\mathbf{c}$ and $\mathbf{H}$, respectively; and since $c_{pm}^* = (\mathbf{c}^+)_{mp}$, it follows from the rules of matrix multiplication that

$$E = \frac{\mathbf{c}^+ \mathbf{H} \mathbf{c}}{\mathbf{c}^+ \mathbf{c}} = \frac{\mathbf{c}^+ (\mathbf{H}_1 + \mathbf{H}_2 + \mathbf{H}_3) \mathbf{c}}{\mathbf{c}^+ \mathbf{c}}. \tag{1.12}$$

On introducing the unit matrix in the form $\mathbf{I} = \mathbf{U}\mathbf{U}^+ = \mathbf{U}^+\mathbf{U}$, where $\mathbf{U}$ is a unitary matrix, and remembering that $(\mathbf{c}\mathbf{U})^+ = \mathbf{U}^+\mathbf{c}^+$, (1.12) becomes

$$E = \frac{\mathbf{c}^+ \mathbf{U}^+ [\mathbf{U}(\mathbf{H}_1 + \mathbf{H}_2 + \mathbf{H}_3)\mathbf{U}^+] \mathbf{U}\mathbf{c}}{(\mathbf{c}^+\mathbf{U}^+)(\mathbf{U}\mathbf{c})}$$

$$= \frac{\mathbf{d}^+ \overline{\mathbf{H}} \mathbf{d}}{\mathbf{d}^+ \mathbf{d}}$$

where $\mathbf{d} = \mathbf{U}\mathbf{c}$.

The last result shows that the energy is unchanged if $\Psi_m'''$ is expanded in terms of the new functions $\Psi_m''$:

$$\Psi_m''' = \sum\limits_{s} d_{sm} \Psi_s''$$

The $\Psi_m''$ are, in fact, related to the single-primed expansion functions through the elements of $\mathbf{U}$:

$$\Psi_m''' = \sum\limits_{p} c_{pm} \Psi_p' = \sum\limits_{q} d_{qm} \Psi_q''$$

$$= \sum\limits_{q} (\mathbf{U}\mathbf{c})_{qm} \Psi_q'' = \sum\limits_{q,\,p} U_{qp} c_{pm} \Psi_q''$$

$$= \sum\limits_{p} c_{pm} \left( \sum\limits_{q} U_{qp} \Psi_q'' \right);$$

that is,

$$\Psi_p' = \sum\limits_{q} U_{qp} \Psi_q''.$$

Thus, although the Hamiltonian matrices look different with respect to the single-primed and double-primed expansion functions the coefficient matrices transform in such a way to keep the energy invariant, a result which is possible only for functions related by a unitary transformation. In the present example, the $L\text{--}S$ and $j\text{--}j$ expansion functions are therefore related by a unitary transformation.

After this preliminary discussion on the handling of approximate wave functions, it is now possible to discuss the construction of the approximate wave functions themselves.

## 1.3.  Determination of approximate electronic wave functions

The molecular orbital and valence bond theories of molecular electronic structure provide the two most widely used methods for obtaining approximate electronic wave functions for polyatomic molecules. The molecular orbital method has achieved the greater popularity because of its apparent computational simplicity. But this has tended to obscure the limitations of the theory, as already discussed in Vol. 1. In practice both methods assume that the many-electron wave function can be built out of a finite number of atomic orbitals centred on the various nuclei in the molecule. The functional form of the atomic orbitals is usually left to individual choice: the most common choice being Slater-type orbitals, with exponents either left as non-linear variation parameters, or fixed by the use of Slater's rules (see Vol. 1). Alternatively, the atomic orbitals may be obtained from the solutions of the appropriate atomic Hartree–Fock equations. The reader is referred to Vol. 1 for a more complete discussion of the general problems concerned with choosing a suitable set of atomic functions.

In spite of the success of the molecular orbital and valence bond theories, the existence of other methods for calculating electronic wave functions, for special kinds of molecule, should not be overlooked. For example, the method of James and Coolidge[7] for treating $H_2$ (already discussed in Vol. 1); the use of bicentric one-electron orbitals in a molecular orbital treatment of homonuclear diatomic molecules;[8] and, finally, an interesting method for hydrides $XH_n$, in which the molecular wave function is constructed from a single set of orbitals centred on the heavy atom X.[9] Unfortunately, all of these methods, although giving accurate molecular wave functions and energies in some cases, are difficult to generalize. For this reason the valence bond and molecular orbital theories, or at least some variant of them, still remain the most useful methods for obtaining approximate electronic wave functions for polyatomic molecules.

The basic aspects of each method have already been discussed in Vol. 1, but it is now necessary to reiterate, and extend, this discussion in preparation for the treatment of polyatomic molecules in Chapters 2 and 3.

## 1.4.  The valence-bond method

In the valence-bond approach to molecular electronic structure, the ground state of a molecule is, in general, represented by a wave function in which the electrons are spin-paired to form core-pairs, bonding-pairs and lone-pairs (if the molecule is a radical, there will be some non-paired electrons). If the spin pairing scheme of lowest energy is not immediately obvious, then, as illustrated by the oxygen molecule, the correct result must be obtained by actual calculation.

The whole approach of the simple valence-bond theory is in direct contradistinction to that of the simplest molecular orbital theory, which is described shortly. For in the latter method, the electron pairs are placed in spatial orbitals extending over the entire molecular framework, rather than in orbitals localized in the same region of space as required in the usual formulation of valence-bond theory. But although the simple valence-bond theory has a direct conceptual appeal, the possibility of performing actual calculations is greatly hindered by the non-orthogonality of the atomic orbitals centred on different nuclei; the net result of which leads to an unwieldy expression for the energy of a polyatomic molecule. This has tended to limit the applications of valence-bond theory to diatomic molecules; but the computational difficulties should become less of a hindrance now, with the advent of larger, faster, computers.

The molecular orbital method does not meet with the same computational difficulties as, for a closed-shell molecule, the basic one-electron functions can be constrained to be orthogonal; and, as shown later, this leads to a simple expression for the molecular electronic energy.

The valence-bond method is now described by illustrating its application to the simple molecules BH and $BH_2^+$; these examples are sufficient to expose the main ideas in the practical application of the theory. The diatomic molecule BH is considered first.

The ground state configuration of the boron atom is $1s^2 2s^2 2p$, and so the unpaired electron in the $2p$ shell can be used for forming an electron pair bond with the electron in the $1s$ orbital, $h$, associated with the hydrogen atom. In this description, there is also a $1s$ core pair and a $2s$ lone pair associated with boron. Thus, following the Heitler–London treatment of $H_2$, as described in Vol. 1, each electron pair function is given as a product of a space function and a spin function, which are symmetric and antisymmetric, respectively, with respect to electron exchange:

$$1s_B(1)1s_B(2).(1/\sqrt{2})\{\alpha(1)\beta(2) - \beta(1)\alpha(2)\}$$
$$2s_B(3)2s_B(4).(1/\sqrt{2})\{\alpha(3)\beta(4) - \beta(3)\alpha(4)\}$$
$$\{2p_B(5)h(6) + 2p_B(6)h(5)\}.(1/\sqrt{2})\{\alpha(5)\beta(6) - \beta(5)\alpha(6)\}.$$

The six-electron molecular wave function is given by the antisymmetrized product of these three two-electron functions:

$$\Psi_0 = N\hat{A}1s_B(1)1s_B(2)2s_B(3)2s_B(4)[2p_B(5)h(6) + h(5)2p_B(6)]$$
$$\times (1/\sqrt{2})[\alpha(1)\beta(2) - \beta(1)\alpha(2)] \times (1/\sqrt{2})[\alpha(3)\beta(4) - \beta(3)\alpha(4)]$$
$$\times (1/\sqrt{2})[\alpha(5)\beta(6) - \beta(5)\alpha(6)]$$
$$= N\hat{A}\phi\Theta \tag{1.13}$$

so ensuring that the overall wave function changes sign under the interchange of *any* pair of electrons. $\phi$, $\Theta$ are the appropriate products of space and spin functions, respectively; $N$ is a constant, usually fixed by the normalization condition, $\int \Psi_0\Psi_0 \, d\tau = 1$, and $\hat{A}$ is the antisymmetrizing operator

$$\sum_{\hat{P}} (-1)^P \hat{P}.$$

The sum over $\hat{P}$ includes all 6! permutations of the electron labels, and $(-1)^P$ gives the parity of the permutation. The operator $\hat{A}$ generates a function of definite permutational symmetry from the simple three-fold product of two-electron functions.

Equation (1.13) is written in the particular form above to show how the Heitler–London treatment of a single electron-pair bond is generalized to deal with several electron-pairs. However, this is not necessarily the best way of writing $\Psi_0$, and some alternative representations are now described.

Since the antisymmetry of the six-electron wave function, with respect to the interchanges $1 \leftrightarrow 2$, $3 \leftrightarrow 4$ and $5 \leftrightarrow 6$, is already incorporated in the two-electron functions, the corresponding interchanges in $\hat{A}$ are redundant. But rather than removing these interchanges from $\hat{A}$, they can be removed from the pair functions without any loss of generality:

$$\Psi_0 = 2^{-3/2}N\hat{A}1s_B(1)1s_B(2)2s_B(3)2s_B(4)[2p_B(5)h(6) + h(5)2p_B(6)] \times$$
$$\times [\alpha(1)\beta(2) - \beta(1)\alpha(2)][\alpha(3)\beta(4) - \beta(3)\alpha(4)][\alpha(5)\beta(6) - \beta(5)\alpha(6)]$$
$$= 2^{-3/2}N\hat{A}(1 - \hat{P}_{12})1s_B(1)1s_B(2)2s_B(3)2s_B(4)[2p_B(5)h(6) + h(5)2p_B(6)] \times$$
$$\times \alpha(1)\beta(2)[\alpha(3)\beta(4) - \beta(3)\alpha(4)][\alpha(5)\beta(6) - \beta(5)\alpha(6)]$$
$$= 2^{-3/2}N\hat{A}(1 - \hat{P}_{12})(1 - \hat{P}_{34})(1 - \hat{P}_{56})1s_B(1)1s_B(2)2s_B(3)2s_B(4) \times$$
$$\times [2p_B(5)h(6) + h(5)2p_B(6)]\alpha(1)\beta(2).\alpha(3)\beta(4).\alpha(5)\beta(6)$$

If $\hat{P}'$ represents any one of the above three interchanges, then

$$\hat{A}\hat{P}' = -\hat{P}'\hat{A} = -\hat{P}'\sum_{\hat{Q}}(-1)^Q\hat{Q} = \sum_{\hat{S}}(-1)^S\hat{S} = \hat{A}$$

because

$$\sum_{\hat{Q}}(-1)^{Q+1}\hat{P}'\hat{Q}$$

is just the sum of the 6! permutations in a different order; and since the interchanges are taken with their parity, each permutation $\hat{S}$ appears with the correct parity factor. For example, if $\hat{Q}$ is an odd permutation, and $-\hat{P}'\hat{Q} = \hat{S}$, then $\hat{S}$ is an even permutation with parity $(-1)^2 = +1$.

Thus,

$$
\begin{aligned}
\Psi_0 &= N'\hat{A}1s_B(1)1s_B(2)2s_B(3)2s_B(4)[2p_B(5)h(6) + h(5)2p_B(6)] \times \\
&\quad \times \alpha(1)\beta(2)\alpha(3)\beta(4)\alpha(5)\beta(6) \\
&= N'\hat{A}1s_B(1)\overline{1s_B}(2)2s_B(3)\overline{2s_B}(4)[2p_B(5)\bar{h}(6) + h(5)\overline{2p_B}(6)] \\
&= N'\{\hat{A}1s_B(1)\overline{1s_B}(2)2s_B(3)\overline{2s_B}(4)2p_B(5)\bar{h}(6) \\
&\quad + \hat{A}1s_B(1)\overline{1s_B}(2)2s_B(3)\overline{2s_B}(4)h(5)\overline{2p_B}(6)\}, \quad (1.14)
\end{aligned}
$$

where the usual notation has been used for spin orbitals: the presence or absence of a bar over a particular orbital indicating $\beta$ or $\alpha$ spin, respectively.

Now each term in (1.14) represents the expansion of a $6 \times 6$ determinant (see Hall[10]), and so $\Psi_0$ can be written in the form

$$
N'\left\{
\begin{vmatrix}
1s_B(1) & 1s_B(2) & \ldots\ldots & 1s_B(6) \\
\overline{1s_B}(1) & \overline{1s_B}(2) & \ldots\ldots & \cdot \\
2s_B(1) & \overline{2s_B}(2) & \ldots\ldots & \cdot \\
\overline{2s_B}(1) & \overline{2s_B}(2) & \ldots\ldots & \cdot \\
2p_B(1) & 2p_B(2) & \ldots\ldots & \cdot \\
h(1) & h(2) & \ldots\ldots & h(6)
\end{vmatrix}
+
\begin{vmatrix}
1s_B(1) & \ldots\ldots & 1s_B(6) \\
\cdot & & \cdot \\
\cdot & & \cdot \\
\cdot & & \cdot \\
h(1) & \ldots\ldots & h(6) \\
\overline{2p_B}(1) & \ldots\ldots & \overline{2p_B}(6)
\end{vmatrix}
\right\}
$$

or, alternatively, in the more compact form

$$
N'\{|1s_B(1)\overline{1s_B}(2)2s_B(3)\overline{2s_B}(4)2p_B(5)\bar{h}(6)| \\
+ |1s_B(1)\overline{1s_B}(2)2s_B(3)\overline{2s_B}(4)h(5)\overline{2p_B}(6)|\}
$$

where the product of the diagonal elements of each determinant is used as a shorthand notation for the complete determinant.

Equation (1.14) therefore shows that a wave function involving one bond pair function is represented by a linear combination of two determinants. The same result is often written symbolically as

$$
\Psi_0 = N'\hat{A}1s_B(1)\overline{1s_B}(2)2s_B(3)\overline{2s_B}(4)\widehat{2p_Bh}
$$

where

$$
\widehat{2p_Bh} = 2p_B(5)h(6)[\alpha(5)\beta(6) - \beta(5)\alpha(6)],
$$

or $[2p_B(5)h(6) + h(5)2p_B(6)]\alpha(5)\beta(6)$.

In the general case of $n$ bond pair functions, the analogous expansion for $\Psi_0$ involves $2^n$ determinants.

In addition to the requirement of antisymmetry, $\Psi_0$ has to display certain other properties. First, in the case of a diatomic molecule, the component of orbital angular momentum along the internuclear axis remains constant, and is therefore a good quantum number. Secondly, in the absence of spin-orbit coupling, $\Psi_0$ must be an eigenfunction of $\hat{S}^2$ and $\hat{S}_z$; the operators corresponding to the square and $z$ component, respectively, of the total spin angular momentum.

Now in the molecular environment the original $(2l + 1)$ fold degeneracy associated with each atomic orbital is lost, and the atomic orbitals are characterized by their component of orbital angular momentum along the internuclear axis. Thus, if the local $z$-axes of the combining atoms are aligned in the direction of the incipient bond then the $1s$, $2s$ and $2p_z$ atomic orbitals in the free atom (all having $m_l = 0$) will have a zero component of angular momentum along the bond in the molecule. The remaining pair of $2p$ atomic orbitals, having $m_l = \pm 1$, will then have components of $\pm 1$ units of angular momentum along the bond. The $s$ and $2p_z$ (or $2p_0$) atomic orbitals are termed $\sigma$ orbitals, while the two remaining $2p$ atomic orbitals are referred to as $\pi$-atomic orbitals. And as the ground state is expected to have a zero component of orbital angular momentum along the bond, the $2p$ atomic orbital used in the construction of $\Psi_0$ can only be a $2p_z$ atomic orbital. $\Psi_0$ therefore represents an approximation to the lowest energy $\Sigma$ state.

The construction of the molecular wave function, as given in (1.13), ensures that $\Psi_0$ is an eigenfunction of $\hat{S}^2$, with zero eigenvalue; that is, $\Psi_0$ corresponds to a singlet spin state. This is readily verified by operating with $\hat{S}^2$ on $\Psi_0$:

$$\hat{S}^2\Psi_0 = (\hat{S}_x^2 + \hat{S}_y^2 + \hat{S}_z^2) = (\hat{S}_+\hat{S}_- + \hat{S}_z^2 - \hat{S}_z)\Psi_0$$

$$= \left[\left(\sum_m \hat{s}_+(m)\right)\left(\sum_n \hat{s}_-(n)\right) + \left(\sum_m \hat{s}_z(m)\right)\left(\sum_n \hat{s}_z(n)\right) - \left(\sum_m \hat{s}_z(m)\right)\right]\Psi_0$$

where $\hat{S}_x, \hat{S}_y$, and $\hat{S}_z$ are operators corresponding to the $x, y$, and $z$ components of the spin angular momentum, respectively; $\hat{S}_\pm = \hat{S}_x \pm i\hat{S}_y$, and $\hat{s}_\pm(m)$, $\hat{s}_z(m)$ are the corresponding operators for electron $m$ (see Vol. 1 and p. 116 for statements of the properties of angular momentum operators).

As seen in Vol. 1, $\hat{S}^2$ and $\hat{A}$ commute, and since

$$\hat{S}^2\Psi_0 = N'\hat{S}^2\hat{A}\phi\Theta = N'\hat{A}\phi(\hat{S}^2\Theta)$$

it is only necessary to deal with the spin part of the wave function explicitly; that is,

$$\hat{S}^2\Psi_0 = N'\hat{A}\phi\{\hat{S}^2[\alpha(1)\beta(2) - \alpha(2)\beta(1)][\alpha(3)\beta(4) - \alpha(4)\beta(3)] \times$$
$$\times \ [\alpha(5)\beta(6) - \alpha(6)\beta(5)]\}$$

$$= N'\hat{A}\phi\left\{\left(\sum_m \hat{s}_+(m)\right)([\beta(1)\beta(2) - \beta(2)\beta(1)][\alpha(3)\beta(4) - \alpha(4)\beta(3)] \times\right.$$
$$\times \ [\alpha(5)\beta(6) - \alpha(6)\beta(5)]$$

$$+ \ [\alpha(1)\beta(2) - \alpha(2)\beta(1)][\beta(3)\beta(4) - \beta(4)\beta(3)]$$
$$\times \ [\alpha(5)\beta(6) - \alpha(6)\beta(5)]$$

$$+ \ [\alpha(1)\beta(2) - \alpha(2)\beta(1)][\alpha(3)\beta(4) - \alpha(4)\beta(3)] \times$$
$$\left. \times \ [\beta(5)\beta(6) - \beta(6)\beta(5)])\right\}.$$

The identical result is also obtained by operating with $\hat{S}^2$ on (1.14). For purposes of clarity, however, it is better to separate the space and spin parts of the molecular wave function: the singlet spin function $\Theta$ is then found to be given very simply by the three-fold product of the appropriate singlet coupled spin pair functions.

In concluding these remarks about the symmetry properties displayed by the ground electronic wave function for BH, it should be noted that $\Psi_0$ is invariant to the symmetry operation of reflection in a plane containing the nuclei: that is, $\Psi_0$ represents an approximation to a molecular $^1\Sigma^+$ electronic state.

The structure represented by $\Psi_0$, which is expected to form a zeroth-order approximation to the lowest energy $^1\Sigma^+$ state, is conventionally termed a covalent structure, as $\Psi_0$ is constructed out of component wave functions pertaining to neutral boron and hydrogen atoms. However, $\Psi_0$ is unlikely to be a good approximate wave function, because (1.13) does not represent the only way of constructing a $^1\Sigma^+$ molecular wave function from appropriate B and H atomic configurations: an improved approximation is obtained by allowing for the interaction between $\Psi_0$ and selected higher energy $^1\Sigma^+$ wave functions. The contribution of these higher energy states is often of crucial importance in valence bond calculations and, for this reason, the construction of such states is now discussed for BH.

In terms of the free B atom many-electron states, (1.13) represents the combining of the $^2P(1s^22s^22p)$ state of B with the $^2S(1s)$ state of H to yield a molecular state of symmetry $^1\Sigma^+$: in this particular instance it is the $M_L = 0$ component of the $^2P$ state which is utilized. Other, higher energy, states of B can be obtained by promoting electrons from the $2s$ to the $2p$ shell. Selected states arising from the $2s2p^2$ and $2p^3$ configurations of B can then couple with the $^2S$ state of H to yield additional $^1\Sigma^+$ molecular states. In the free atom, however, no coupling can occur between states arising from these excited configurations, as the respective sets of terms are of different parity (even and odd number of $p$ electrons respectively). It is the breakdown of the atomic spherical symmetry which allows the interaction to occur in the molecule. For the purposes of the present discussion, though, only those molecular states are considered which arise from the $2s2p_z^2$ configuration of boron: other states, involving, for example, the occupancy of the boron $2p_{\pm 1}$ atomic orbitals, must be considered in a more complete treatment.

The structure involving a bond pair between $2s_B$ and $h$, with a lone-pair in the boron $2p_z$ atomic orbital, is represented by

$$\Psi_1 = N\hat{A} 1s_B(1)\overline{1s_B}(2) 2p_B(3)\overline{2p_B}(4)\widehat{2s_B h}$$

An improved pair of $^1\Sigma^+$ wave functions is therefore obtained by expressing the wave function in the form

$$\Psi'_M = c_{0M}\Psi_0 + c_{1M}\Psi_1$$

and minimizing the energy with respect to $c_{0M}$ and $c_{1M}$. This leads in the usual way to the solution of the secular determinant

$$\begin{vmatrix} H_{00} - ES_{00} & H_{01} - ES_{01} \\ H_{10} - ES_{10} & H_{11} - ES_{11} \end{vmatrix} = 0 \qquad (1.15)$$

where $H_{ij} = \int \Psi_i \hat{H} \Psi_j \, d\tau$ and $S_{ij} = \int \Psi_i \Psi_j \, d\tau$. The lowest root of (1.15) corresponds to the improved ground state energy, and the eigenvector components, $c_{i0}$, are determined by the solution of the appropriate secular equations as described in Vol. 1.

Since $\Psi_0$ contains a $2s$ lone-pair and $\Psi_1$ a $2p_z$ lone-pair, the improved approximation to the ground state wave function will contain a lone pair with $2s$ and partial $2p_z$ character. Similarly, the bonding pair will be described in terms of some mixture of B $2s$ and $2p$ atomic orbitals. This mixing of the $2s$ and $2p_z$ atomic functions is a direct consequence of the molecular environment, and can be regarded as a form of polarization of B by H; that is, the shape of the charge distribution around B is changed by the presence of hydrogen. Alternatively, this polarization can be recognized from the start of the calculation by expressing the approximate wave functions in terms of hybrid atomic orbitals, rather than ordinary atomic orbitals. Since the hybrid atomic orbitals are strongly directional in character, the lone-pair and bonding electrons on the same atom can then be localized in different regions of space in a chemically suggestive manner. In the present instance, for example, the $2s$ and $2p_z$ atomic orbitals of boron can be replaced by the orthogonal combinations

$$di_1 = (1/\sqrt{2})(2s_B + 2p_B)$$
$$di_2 = (1/\sqrt{2})(-2s_B + 2p_B) \qquad (1.16)$$

which have their maximum amplitudes in regions normally ascribed to the bonding and lone-pair electrons, respectively (the local $z$-axis at boron is directed towards hydrogen).

This discussion suggests, therefore, that it may be possible to use the chemically motivated structure

$$\Psi_0' = N'' \hat{A} 1s_B(1)\overline{1s_B}(2) \, di_2(3) \, \overline{di_2}(4) \, \widehat{di_1 h} \qquad (1.17)$$

for describing the electronic ground state of BH. However, a wave function of this form requires handling with extreme caution, as the expansion of (1.17), with the repeated use of (1.16), yields

$$\Psi_0' = N'' \{ | 1s_B(1)\overline{1s_B}(2) \, di_2(3) \, \overline{di_2}(4) \, di_1(5)\overline{h}(6) |$$
$$+ | 1s_B(1)\overline{1s_B}(2) \, di_2(3) \, \overline{di_2}(4) h(5) \, \overline{di_1}(6) | \}$$

$$= 2^{-1/2} N'' \{ |1s_B(1)\overline{1s_B}(2)2s_B(3)\overline{2s_B}(4)2p_B(5)h(6)|$$

$$+ |1s_B(1)\overline{1s_B}(2)2s_B(3)\overline{2s_B}(4)h(5)\overline{2p_B}(6)|$$

$$+ |1s_B(1)\overline{1s_B}(2)2p_B(3)\overline{2p_B}(4)2s_B(5)h(6)|$$

$$+ |1s_B(1)\overline{1s_B}(2)2p_B(3)\overline{2p_B}(4)h(5)\overline{2s_B}(6)| \}$$

$$= \text{const.} \left\{ \Psi_0 + \left(\frac{N'}{N}\right) \Psi_1 \right\}$$

showing that the use of (1.17) corresponds to working with an expansion for $\Psi'_0$ in which the relative weights of $\Psi_0$ and $\Psi_1$ are fixed by the choice of hybrids, rather than by energy minimization. The necessary flexibility can be introduced only by considering the contribution of the long-bonded structure

$$\Psi'_1 = N''' \hat{A} 1s_B(1)\overline{1s_B}(2) \, \mathrm{d}i_1(3) \, \overline{\mathrm{d}i_1}(4) \, \widehat{\mathrm{d}i_2 h} = \text{const.} \left\{ \Psi_0 - \left(\frac{N'}{N}\right) \Psi_1 \right\}$$

which, on expansion in terms of structures based on ordinary atomic orbitals, is seen to consist of a different linear combination of $\Psi_0$ and $\Psi_1$. A linear combination of $\Psi'_0$ and $\Psi'_1$, with one parameter to be determined by energy minimization (the other is determined by normalization), will therefore yield the same energy as the calculation based on the use of ordinary atomic orbitals.

Thus, in this simple example, it is irrelevant whether the energy calculation is based on the admixture of the structures $\Psi_0$, $\Psi_1$ or the structures $\Psi'_0$, $\Psi'_1$. On the other hand, for polyatomic molecules, where it is impossible to consider all feasible structures, the approach using hybrid atomic orbitals is very appealing. For, in general, it is not very difficult to list reasonable structures on the basis of chemical intuition; but more will be said about this particular problem later in this section.

As far as BH is concerned, all the structures considered so far can be described as covalent in nature. In addition, only those structures have been considered which involve a rearrangement of the electrons within the L-shell of boron. It might be thought that new, though energetically unfavourable, structures could be formed by uncoupling the core electrons to form structures like

$$N' \hat{A} \widehat{1s_B h} \widehat{1s_B 2p_B} 2s_B(5)\overline{2s_B}(6)$$

However, the expansion of this structure in terms of determinants (two of which vanish as they both have two identical rows) just yields $-\Psi_0$; so nothing is gained by trying to uncouple the spins of doubly occupied orbitals, providing all structures involving the singly occupied orbitals are considered. New covalent structures can be formed, though, by exciting a core electron into the L-shell of boron. But the contribution of these structures, to the ground-state wave function, is expected to be small because of the large excitation energy required. Structures of this kind are therefore ignored in the present discussion.

Now in BH the cylindrically symmetrical potential, arising from the (fixed) nuclei, will cause a readjustment of the electron distribution so that the correct balance is maintained between the average kinetic and average potential energies (see Vol. 1 for a discussion of the virial theorem). This readjustment is usually envisaged in terms of a change in orbital exponents (change in orbital size), accompanied by a general transfer of charge from one region of the molecule to another. Both effects can in principle be described through the addition of higher energy atomic functions to the original set of atomic orbitals; for example, in the case of BH, by the addition of $3s$, $3p$, $3d$, ... atomic orbitals on B and $2s$, $2p$, ... atomic orbitals on H. In practice the increase in number of the atomic orbitals leads to further problems; one of which is concerned with the handling of the vastly increased number of structures. However, a slight gain is made because the need for optimizing valence orbital exponents becomes less critical when using an extended basis set; for the additional functions themselves can account for most of the changes in orbital size (the reader is referred to Vol. 1 for a discussion of the properties of complete sets of functions). Unfortunately, the use of extended sets of atomic orbitals results in the wave function becoming a complex mathematical entity, devoid of any simple intuitive meaning. For this reason, it is often thought more desirable to describe distortions of the charge distribution in terms of "ionic" structures, constructed from the valence shell set of atomic orbitals; and any further readjustment of the charge distribution is then obtained by optimizing the limited number of atomic orbital exponents. The designation of structures as "covalent" or "ionic" has an obvious chemical connotation; but this must not be taken too literally, as the non-orthogonality of the corresponding wave functions precludes any unique identity of structure type. The terms "covalent" and "ionic" are therefore used only in a bookkeeping sense, and any objective discussion of the charge distribution must be made in terms of the electron density function which is discussed in Section 1.7.

Typical ionic structures for BH, formally describing the drift of charge from boron to hydrogen, are of the form

$$N\hat{A}1s_{\mathrm{B}}(1)\overline{1s_{\mathrm{B}}}(2)2s_{\mathrm{B}}(3)\overline{2s_{\mathrm{B}}}(4)h(5)\bar{h}(6); \quad N\hat{A}1s_{\mathrm{B}}(1)\overline{1s_{\mathrm{B}}}(2)\widehat{2s_{\mathrm{B}}2p_{\mathrm{B}}}h(5)\bar{h}(6)$$

where, on average, two valence electrons are now associated with both boron and hydrogen. Charge transfer in the opposite sense is described by the structure

$$N\hat{A}1s_{\mathrm{B}}(1)\overline{1s_{\mathrm{B}}}(2)2s_{\mathrm{B}}(3)\overline{2s_{\mathrm{B}}}(4)2p_{\mathrm{B}}(5)\overline{2p_{\mathrm{B}}}(6).$$

There are, of course, ionic structures of both kinds which involve occupancy of the $2p_{\pm 1}$ orbitals, but these are disregarded in the present discussion.

Ionic structures can also be constructed in terms of hybrid atomic orbitals (1.16); but, for complete flexibility, structures corresponding to all feasible orbital occupancies must be considered.

An improved approximation to the $^1\Sigma^+$ molecular ground state wave function is therefore obtained by considering the participation of all reasonable covalent and ionic structures, say $n$ in number:

$$\Psi'_0 = \sum_{p=1}^{n} c_{p0}\Psi'_p$$

The calculation of the optimum $c_{p0}$ then proceeds in the usual way, once the values for the matrix elements $H_{ij}$ and $S_{ij}$ are known. These matrix elements can be obtained, in a rather tedious manner, by expanding $\Psi_i$ and $\Psi_j$ in terms of their component determinantal wave functions, and then using the basic formulae (see Appendix I) for the matrix elements of one- and two-electron operators (and the unit operator) between determinantal wave functions constructed from non-orthogonal one-electron functions. The reader is referred to the recent literature for more efficient methods of evaluating these matrix elements.[11–12]

The above discussion has highlighted a number of problems occurring in the application of valence bond theory to diatomic molecules containing more than two electrons. The treatment of polyatomic molecules is much more involved, as, apart from the problem of atomic orbital non-orthogonality, it is often difficult to enumerate the correct number of linearly independent structures arising from the given electron configurations of the combining atoms. Even in the simplest case of BH, considered above, it was shown how an apparently new pairing scheme was in fact a disguised form of one of the pairing schemes already considered. The problem is much more acute for polyatomic molecules, and it therefore requires a systematic method for predicting the correct number of independent pairing schemes. But before discussing how this is accomplished, the basic ideas will be introduced in a simple way by looking at some of the pairing schemes appropriate to the linear configuration of the molecule $BH_2^+$.

Structures can be envisaged which are based on either $B^+$ and two hydrogen atoms, or B and one hydrogen atom and one proton. For purposes of illustration, though, only those structures are considered which involve the valence electron configurations $2s^2$, $2s2p_z$, and $2p_z^2$ of $B^+$; and any structures involving excitation of core electrons, or occupancy of $2p_{\pm 1}$ atomic orbitals of boron, are disregarded.

The $2s^2$ and $2p_z^2$ configurations of $B^+$ both give rise to structures in which the electrons in the two hydrogen $1s$ atomic orbitals are coupled together:

$$\Psi_0 = \hat{A}1s_B(1)\overline{1s_B}(2)2s_B(3)\overline{2s_B}(4)\widehat{h_1 h_2}$$

$$= \hat{A}1s_B(1)\overline{1s_B}(2)2s_B(3)\overline{2s_B}(4)[h_1(5)\overline{h}_2(6) - \overline{h}_1(5)h_2(6)]/\sqrt{2}$$

$$\Psi_1 = \hat{A}1s_B(1)\overline{1s_B}(2)2p_B(3)\overline{2p_B}(4)\widehat{h_1 h_2}$$

B

(normalizing constants are neglected for simplicity). No new structures are gained by uncoupling the $2s$ or $2p_z$ electron pairs.

The $2s2p_z$ configuration of $B^+$ gives rise to several structures. For example, $2s_B$ can be paired with $h_1$ and $2p_B$ with $h_2$; or $2s_B$ can be paired with $h_2$ and $2p_B$ with $h_1$; in addition, a long-bonded structure can be formulated which involves the pairing of $2s_B$ with $2p_B$ and $h_1$ with $h_2$. The corresponding structure wave functions are denoted by $\Psi_2$, $\Psi_3$ and $\Psi_4$, respectively, and have the forms:

$$\Psi_2 = \hat{A}1s_B(1)\overline{1s_B}(2)\overParen{2s_Bh_1}\,\overParen{2p_Bh_2}$$
$$= \hat{A}1s_B(1)\overline{1s_B}(2)2s_B(3)h_1(4)2p_B(5)h_2(6) \times$$
$$\times\ (1/\sqrt{2})[\alpha(3)\beta(4) - \beta(3)\alpha(4)].(1/\sqrt{2})[\alpha(5)\beta(6) - \beta(5)\alpha(6)]$$
$$\Psi_3 = \hat{A}1s_B(1)\overline{1s_B}(2)2s_B(3)h_2(4)2p_B(5)h_1(6).(1/\sqrt{2})[\alpha(3)\beta(4) - \beta(3)\alpha(4)]$$
$$\times\ (1/\sqrt{2})[\alpha(5)\beta(6) - \beta(5)\alpha(6)]$$
$$\Psi_4 = \hat{A}1s_B(1)\overline{1s_B}(2)2s_B(3)2p_B(4)h_1(5)h_2(6).(1/\sqrt{2})[\alpha(3)\beta(4) - \beta(3)\alpha(4)]$$
$$\times\ (1/\sqrt{2})[\alpha(5)\beta(6) - \beta(5)\alpha(6)]. \tag{1.18}$$

The ordering of the atomic orbitals in each $\Psi_i$ is significant, and any reordering is best performed by expanding the appropriate $\Psi_i$ as a sum of four determinants; then interchanging rows, and finally rewriting the result in the form of (1.18). If the same sequence of atomic orbitals is required for each $\Psi_i$, for example $2s2ph_1h_2$, then the spin functions change their appearance:

$$\Psi_2 = \hat{A}1s_B(1)\overline{1s_B}(2)2s_B(3)h_1(4)2p_B(5)h_2(6) \times$$
$$\times\ [\tfrac{1}{2}(\alpha(3)\beta(4)\alpha(5)\beta(6) - \alpha(3)\beta(4)\beta(5)\alpha(6) - \beta(3)\alpha(4)\alpha(5)\beta(6)$$
$$+\ \beta(3)\alpha(4)\beta(5)\alpha(6))]$$
$$= \hat{A}1s_B(1)\overline{1s_B}(2)2s_B(3)2p_B(4)h_1(5)h_2(6) \times$$
$$\times\ [\tfrac{1}{2}(-\ \alpha(3)\alpha(4)\beta(5)\beta(6) + \alpha(3)\beta(4)\beta(5)\alpha(6)$$
$$+\ \beta(3)\alpha(4)\alpha(5)\beta(6) - \beta(3)\beta(4)\alpha(5)\alpha(6))],$$
$$\Psi_3 = \hat{A}1s_B(1)\overline{1s_B}(2)2s_B(3)2p_B(4)h_1(5)h_2(6) \times$$
$$\times\ [\tfrac{1}{2}(\alpha(3)\alpha(4)\beta(5)\beta(6) - \alpha(3)\beta(4)\alpha(5)\beta(6)$$
$$-\ \beta(3)\alpha(4)\beta(5)\alpha(6) + \beta(3)\beta(4)\alpha(5)\alpha(6))],$$
$$\Psi_4 = \hat{A}1s_B(1)\overline{1s_B}(2)2s_B(3)2p_B(4)h_1(5)h_2(6) \times$$
$$\times\ [\tfrac{1}{2}(\alpha(3)\beta(4)\alpha(5)\beta(6) - \alpha(3)\beta(4)\beta(5)\alpha(6)$$
$$-\ \beta(3)\alpha(4)\alpha(5)\beta(6) + \beta(3)\alpha(4)\beta(5)\alpha(6))]. \tag{1.18'}$$

The wave functions corresponding to the three structures can therefore be written in the compact form

$$\Psi_i = \hat{A}1s_B(1)\overline{1s_B}(2)2s_B(3)2p_B(4)h_1(5)h_2(6)\Theta_i$$

where the $\Theta_i$ are the appropriate spin functions in square brackets in (1.18').

But as $\Theta_3$ is given by $-\Theta_4 - \Theta_2$, one of the $\Psi_i$, for example $\Psi_2$, is redundant: that is, it is possible to construct only two singlet states from a configuration of four singly occupied atomic orbitals.

The spin functions associated with the two chosen singlet states are linearly independent, but not orthogonal:

$$S_{34} = \int \Theta_3 \Theta_4 \, ds_3 \, ds_4 \, ds_5 \, ds_6 = -\tfrac{1}{2}.$$

It is conventional, however, to work with spin functions which are mutually exclusive: that is, there is no contamination of $\Theta_3$, by $\Theta_4$, and vice versa. The appropriate orthogonalization can be accomplished without any loss of generality, by noting that the total electronic wave function contains the following contribution from the two singlet structures under consideration:

$$\Psi_0' = \ldots + c_{30} \hat{A} \phi \Theta_3 + c_{40} \hat{A} \phi \Theta_4 + \ldots$$

On adding and subtracting a small amount of $\Psi_4$, and rearranging terms, $\Psi_0'$ can be rewritten in the form

$$\ldots + c_{30} \hat{A} \phi (\Theta_3 + \lambda \Theta_4) + (c_{40} - c_{30}\lambda) \hat{A} \phi \Theta_4 + \ldots$$

The value of $\lambda$ is now chosen so that the spin function $\Theta_3 + \lambda \Theta_4$ is orthogonal to $\Theta_4$:

$$\Psi_0' = \ldots + c_{30}' \hat{A} \phi \frac{(\Theta_3 - S_{34}\Theta_4)}{\sqrt{(1 - S_{34}^2)}} + c_{40}' \hat{A} \phi \Theta_4 + \ldots$$

$$= \ldots + c_{30}' \hat{A} \phi \Theta_3' + c_{40}' \hat{A} \phi \Theta_4 + \ldots$$

and $c_{30}'$ is defined so that $\Theta_3'$ is normalized. The actual form for $\Theta_3'$, in terms of the $\alpha$- and $\beta$-spin functions, follows directly from the definitions of $\Theta_3$, $\Theta_4$ and the value of $S_{34}$:

$$\Theta_3' = (2/\sqrt{3})(\Theta_3 + \tfrac{1}{2}\Theta_4)$$

$$= (1/\sqrt{12})\{2[\alpha(3)\alpha(4)\beta(5)\beta(6)] + 2[\beta(3)\beta(4)\alpha(5)\alpha(6)] - \alpha(3)\beta(4)\alpha(5)\beta(6)$$
$$- \beta(3)\alpha(4)\beta(5)\alpha(6) - \alpha(3)\beta(4)\beta(5)\alpha(6) - \beta(3)\alpha(4)\alpha(5)\beta(6)\}.$$

The spin functions $\Theta_3$, $\Theta_4$ (dropping the prime) therefore represent a possible pair of orthogonal singlet spin functions appropriate to the four spin problem:

$$\hat{S}^2 \Theta_i = 0 \qquad i = 3, 4.$$

There now remains the problem of the spatial symmetry displayed by the ground state wave function. As far as closed shell molecules are concerned, the ground state will be characterized by a totally symmetric singlet wave function: that is, the wave function is left unaltered by any operation of the molecular point group, $G$ (here $D_{\infty h}$). However, the electronic states of molecules possessing the covering group $D_{\infty h}$ are classified further as even or odd ($g$ or $u$, respectively) according to their behaviour under the operation of inversion, $\hat{J}$. Although a more detailed discussion of the effects of symmetry operations on atomic orbitals is left until the next chapter, it is not difficult to see that the operation of inversion induces the orbital transformations

$$h_1 \leftrightarrow h_2; \quad 2s_{\mathrm{B}} \rightarrow 2s_{\mathrm{B}}; \quad 2p_{\mathrm{B}} \rightarrow -2p_{\mathrm{B}}$$

Hence, the effect of $\hat{J}$ on $\Psi_4$ is given by

$$\hat{J}\Psi_4 = -\hat{A}1s_{\mathrm{B}}(1)\overline{1s_{\mathrm{B}}}(2)2s_{\mathrm{B}}(3)2p_{\mathrm{B}}(4)h_2(5)h_1(6)\Theta_4$$

which, on rearrangement, becomes

$$\hat{A}1s_{\mathrm{B}}(1)\overline{1s_{\mathrm{B}}}(2)2s_{\mathrm{B}}(3)2p_{\mathrm{B}}(4)h_1(5)h_2(6) \times$$
$$\times \tfrac{1}{2}\{\alpha(3)\beta(4)\beta(5)\alpha(6) - \alpha(3)\beta(4)\alpha(5)\beta(6)$$
$$- \beta(3)\alpha(4)\beta(5)\alpha(6) + \beta(3)\alpha(4)\alpha(5)\beta(6)\} = -\Psi_4.$$

Each of the remaining structures $\Psi_0$, $\Psi_1$, and $\Psi_3$ is invariant under $\hat{J}$, and so they all contribute to the lowest energy $^1\Sigma_g^+$ state.

It should be noted that in this particular instance, the four structures, based on electron configurations of $\mathrm{B}^+$, are already adapted to the symmetry of the molecule. If this were not the case, then it would be necessary to use the operator (see footnote p. 69)

$$\sum_{\hat{R}} \chi_\Gamma^*(\hat{R})\hat{R}$$

in order to construct suitable linear combinations of structure wave functions transforming like the irreducible representation $\Gamma$ of $G$ (for ground states this is usually the totally symmetric irreducible representation). In the general case, it may also be necessary to apply the operator several times until the required number of (linearly independent) symmetry adapted wave functions is found.

The above discussion can also be developed in terms of structures based on the use of hybrid atomic orbitals, rather than the use of the $2s$ and $2p_z$ atomic orbitals of boron. In this approach, the structures involving $\mathrm{d}i_1^2$ and $\mathrm{d}i_2^2$ configurations of $\mathrm{B}^+$ each give rise to one singlet coupled wave function (compare with the $2s^2$ and $2p_z^2$ configurations); while the configuration $\mathrm{d}i_1\mathrm{d}i_2$ gives rise to two singlet states, exactly paralleling the case of the $2s2p_z$ configuration:

$$\Psi_0' = \hat{A}1s_{\mathrm{B}}(1)\overline{1s_{\mathrm{B}}}(2)\, \mathrm{d}i_1(3)\, \mathrm{d}i_1(4)\, h_1(5)\, h_2(6)\Theta_4 = [\Psi_0 + \sqrt{2}.\Psi_4 + \Psi_1]/\sqrt{2},$$

$$\Psi_1' = \hat{A}1s_{\mathrm{B}}(1)\overline{1s_{\mathrm{B}}}(2)\, \mathrm{d}i_2(3)\, \mathrm{d}i_2(4)\, h_1(5)h_2(6)\Theta_4 = [\Psi_0 - \sqrt{2}.\Psi_4 + \Psi_1]/\sqrt{2},$$

$$\Psi_2' = \hat{A}1s_{\mathrm{B}}(1)\overline{1s_{\mathrm{B}}}(2)\, \mathrm{d}i_1(3)\, \mathrm{d}i_2(4)h_1(5)h_2(6)\Theta_4 = -[\Psi_0 - \Psi_1]/\sqrt{2},$$

$$\Psi_3' = \hat{A}1s_{\mathrm{B}}(1)\overline{1s_{\mathrm{B}}}(2)\, \mathrm{d}i_1(3)\, \mathrm{d}i_2(4)h_1(5)h_2(6)\Theta_3 = \Psi_3.$$

The wave functions are also given in terms of structures defined with respect to ordinary atomic orbitals.†

† Notice that the inclusion of normalizing constants alters these formulae slightly, but this is not important here.

$\Psi'_2$, $\Psi'_3$ are already totally symmetric singlet state wave functions; but, under the operation of inversion, $\Psi'_0 \rightarrow \Psi'_1$. And, in this particular case, it is not difficult to see that the linear combinations $\Psi'_0 + \Psi'_1$ and $\Psi'_0 - \Psi'_1$ are eigenfunctions of $\hat{J}$, with eigenvalues $+1$ and $-1$, respectively. The lowest energy $^1\Sigma_g^+$ state therefore contains contributions from the three totally symmetric singlet states $\Psi'_2$, $\Psi'_3$ and $\Psi'_0 + \Psi'_1$ (unnormalized).

An entirely equivalent set of wave functions can be obtained from the spin pairing schemes

$$\Psi''_0 = \hat{A} 1s_B(1)\overline{1s_B}(2)\, di_1(3)\, \overline{di_1}(4)\widehat{h_1 h_2} = \Psi'_0/\sqrt{2}$$

$$\Psi''_1 = \hat{A} 1s_B(1)\overline{1s_B}(2)\, di_2(3)\, \overline{di_2}(4)\widehat{h_1 h_2} = \Psi'_1/\sqrt{2}$$

$$\Psi''_2 = \hat{A} 1s_B(1)\overline{1s_B}(2)\, \widehat{di_1 h_1}\, \widehat{di_2 h_2} = -(\sqrt{3}/2)\Psi'_3 - \tfrac{1}{2}\Psi'_2$$

$$\Psi''_3 = \hat{A} 1s_B(1)\overline{1s_B}(2)\, \widehat{di_1 h_2}\, \widehat{di_2 h_1} = (\sqrt{3}/2)\Psi'_3 - \tfrac{1}{2}\Psi'_2$$

without paying attention to the orthogonality of the spin functions. $\Psi''_2$ is the chemically motivated structure representing the formation of two electron-pair bonds, within the approximation of "perfect pairing". The expansion of $\Psi''_2$, in terms of structures involving ordinary atomic orbitals, is very interesting, as all basic structures appear, but with weights fixed by the choice of the hybrid atomic orbitals, rather than by energy minimization. It is the contribution of the three remaining structures, corresponding to spin pairings which are considered unlikely from a chemically intuitive point of view, which ensures that the total wave function has the necessary flexibility.

In the two examples considered above, it is not difficult to find the correct number of totally symmetric singlet wave functions. In general, however, it is not at all obvious how to enumerate the actual number of linearly independent totally symmetric singlet states; and the example of methane is now used to illustrate some further aspects of the problem (molecules possessing a ground state with a different spatial symmetry and spin multiplicity can be dealt with in a similar way).

Consider the covalent structures only, based on either ordinary or hybrid atomic orbitals. There are structures in which (a) all carbon orbitals are singly occupied, (b) one carbon orbital is doubly occupied and two are singly occupied, and (c) two carbon orbitals are doubly occupied; (a) corresponds to an eight-spin system (four unpaired spins on both the carbon and hydrogen atoms initially), and it is not immediately obvious how many totally symmetric singlet states can arise; (b) is a six-spin system, and there are twelve ways of choosing each configuration; (c) is a four-spin system, and each configuration can arise in one of six ways. As will be proved shortly, there are fourteen singlets associated with the (a) configuration; five singlets with (b) and two singlets with (c). Thus, there are $14 + 5 \times 12 + 2 \times 6 = 86$ singlet structures, forming the basis for a reducible representation of $T_d$. The reduction of this representation, by standard group theoretical techniques, shows that there are only seven totally symmetric singlet structures. But it must be remembered that these seven structures arise from only the limited number of covalent structures under consideration—if ionic structures involving the species $C^+$, $C^-$, $H^+$, $H^-$ are included, then the number of totally symmetric singlet states increases by sixty-nine.

The method of calculating the maximum number of singlet states which can arise from a given number of unpaired spins is now described.

The $z$-component of spin angular momentum, $M_S$, must be zero for a singlet $N$-electron state: this requires each $N$-fold product of spin functions to contain the same number of $\alpha$- and $\beta$-spin functions. The easiest way of counting the number of singlets is to start at the largest permissible value of $S$, and work down through successive values until $S = 0$ is reached. The reason for this is that of all the possible $^NC_{N/2}$ products having $M_S = 0$, some combinations of these products belong to $M_S = 0$ components of multiplets with $S \neq 0$. Hence, it is first necessary to find the number of these higher spin multiplets.

One component of the state of highest spin multiplicity arises when $S = M_S = N/2$: that is, all spin functions in the $N$-fold product are $\alpha$ spin functions. There is only one way ($^NC_0$) of achieving this distribution. The number of states with $M_S = N/2 - 1$ is equal to the number of ways of forming an $N$-fold product of one $\beta$-spin function and $N - 1$ $\alpha$-spin functions, i.e. $^NC_1$ ways; so the number of states with $S = N/2 - 1$ is given by $^NC_1 - {}^NC_0$. By continuing this argument, the number of states with $S = N/2 - m$ is found to be

$$^NC_m - ({}^NC_{m-1} - {}^NC_{m-2}) - ({}^NC_{m-2} - {}^NC_{m-3}) \ldots - {}^NC_0$$
$$= {}^NC_m - {}^NC_{m-1}.$$

The number of singlet states, for which $m = N/2$, is therefore given by

$$^NC_{N/2} - {}^NC_{N/2-1} = \frac{N!}{(N/2)!(N/2 + 1)!}. \tag{1.19}$$

Hence, in the above example of methane, each (a)- and (b)-type configuration, characterized by $N = 8$ and $N = 6$, respectively, gives rise to fourteen and five singlet states, respectively.

The remaining problem of determining the allowed number of totally symmetric singlet functions is best solved—at least for molecules possessing a high degree of symmetry—by using the properties of the symmetric group; and the reader is referred to the literature for further details[13-14] (see also Bibliography). However, the number of such structures can also be obtained, though less elegantly, by resolving the reducible representation in the usual way. The actual symmetry adapted functions themselves are found by applying the operator

$$\sum_{\hat{R}} \chi_\Gamma^*(\hat{R})\hat{R}$$

to the various structures, until the required number of linearly independent states has been found.

For a polyatomic molecule, the possibility of performing anything like a complete valence-bond calculation is very remote: the large number of structures to be considered creates a new problem, in addition to the one of atomic orbital non-orthogonality. To make the calculations tractable, it is clearly necessary to restrict the number of structures in some way—for example, by selecting only chemically feasible structures, based on the use of hybrid atomic orbitals. Although this procedure may sometimes present problems—particularly when there is valence-shell expansion, as discussed in the next chapter—it often provides a useful method for obtaining reasonable molecular wave functions. For example, in a valence-bond calculation on methane, the ground-state wave function could be approximated by the totally symmetric singlet structure shown in Fig. 1.1, which is based on the

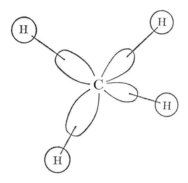

Fig. 1.1. One of the three totally symmetric singlet covalent structures (the one corresponding to "perfect pairing") for methane, arising from singly occupied carbon $sp^3$ hybrid atomic orbitals and hydrogen $1s$ atomic orbitals. Interatomic electron pairing is indicated by a line joining singly occupied orbitals.

use of singly occupied tetrahedrally oriented hybrid atomic orbitals of carbon. On expanding the hybrid orbitals in terms of ordinary atomic orbitals, the wave function is found to contain contributions from covalent structures based on the $2s^2 2p^2$, $2s 2p^3$ and $2p^4$ valence electron configurations of carbon; the relative weights of which are determined by the choice of hybrid orbitals. It is possible, therefore, that too much emphasis may be placed on structures involving the higher energy configurations of carbon. This is just one of the limitations that has to be accepted if the molecular wave function is approximated by only one of the seven totally symmetric singlet covalent structures arising from all possible hybrid orbital occupancies of neutral carbon (three structures arise from singly occupied orbitals, and four from doubly occupied orbitals). However, in this instance, where the terms arising from the three configurations are not too widely separated in energy, and the $2p$ shell is at least occupied in the ground-state configuration, the implicit inclusion of the higher energy structures with fixed weights may not be too serious a disadvantage. It is perhaps necessary to accept some limitation of this kind to enable calculations to be performed.

Although the balance of covalent structures is fixed by the assumption of the perfect pairing approximation—at least when hybrid orbitals are used— some flexibility can be introduced by considering the admixture of chemically feasible ionic structures of the kind shown in Fig. 1.2. Notice that neither of these structures is symmetry adapted: each one is merely a representative member of a set of four equivalent structures which must be considered collectively.

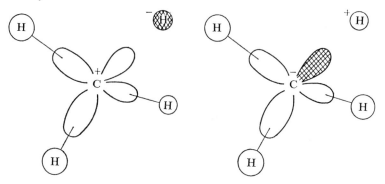

Fig. 1.2. Possible ionic structures contributing to the ground electronic state of methane. Interatomic electron pairing is indicated by a line joining singly occupied orbitals, and orbitals containing lone-pairs of electrons are indicated by crossed hatching. All remaining orbitals are unoccupied.

The selection of only a limited number of reasonable totally symmetric singlet structures, of both covalent and ionic type, at least retains the concept of the chemical bond in outline. The additional flexibility provided by the inclusion of the ionic structures enables the electron distribution in each bond to be determined optimally. Thus, neglecting $1s$ core pairs,

$$\Psi_0' = c_{00}\hat{A}\widehat{te_1h_1}\widehat{te_2h_2}\widehat{te_3h_3}\widehat{te_4h_4}$$
$$+ c_{10}[\hat{A}\widehat{te_1\bar{te}_1}\widehat{te_2h_2}\widehat{te_3h_3}\widehat{te_4h_4} + \hat{A}\widehat{te_1h_1}\widehat{te_2\bar{te}_2}\widehat{te_3h_3}\widehat{te_4h_4} + \ldots]$$
$$+ c_{20}[\hat{A}\widehat{h_1\bar{h}_1}\widehat{te_2h_2}\widehat{te_3h_3}\widehat{te_4h_4} + \ldots]$$

where $te_i$ is the carbon $sp^3$ hybrid atomic orbital directed towards $h_i$. Multiply ionic structures can also be considered if necessary.

Analogous assumptions allow the water molecule to be treated in terms of structures like those shown in Fig. 1.3. Apart from the linear coefficients determining the relative weights of the various structures, the hybrid atomic

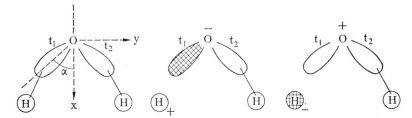

FIG. 1.3. Possible covalent and ionic structures contributing to the ground electronic state of the water molecule. Interatomic electron pairing is indicated by a line joining singly occupied orbitals, and orbitals containing lone-pairs of electrons are indicated by crossed hatching. All remaining orbitals are unoccupied.

orbitals on oxygen are undetermined to within a single parameter, $\lambda$: the square of which gives the amount of $2s$ character in each of the bonding hybrids. This additional parameter arises because symmetry requires only that the hybrids $t_1$ and $t_2$ (see Fig. 1.3) have the same ratio of $2s$ to $2p$ character; a similar condition holding for the lone pair hybrids. Thus, the molecular symmetry does not determine the apportioning of the total $2s$ character between bonding and lone pair hybrids, in direct contrast to the situation in methane where the forms of the tetrahedral hybrid orbitals are completely determined by symmetry.

With the axis system as defined in Fig. 1.3, it is not difficult to show that the four hybrids have the forms

$$t_1 = \lambda 2s - (1/\sqrt{2})2p_y + \mu 2p_x; \quad t_2 = \lambda 2s + (1/\sqrt{2})2p_y + \mu 2p_x$$
$$t_3 = \mu 2s + (1/\sqrt{2})2p_z - \lambda 2p_x; \quad t_4 = \mu 2s - (1/\sqrt{2})2p_z - \lambda 2p_x$$
$$(\mu^2 + \lambda^2 = \tfrac{1}{2}, \quad \mu = (1/\sqrt{2}) \cot \alpha)$$

where, as just discussed, the apportioning of the $2s$ character between bonding and lone pair functions must be determined by energy minimization. If the hybrid orbitals are assumed to be directed along the O–H bonds then the apportioning of $2s$ character is fixed by the molecular geometry, and one of the variation parameters is removed from the problem.

In the above treatment of valence bond theory, the number of structures has been reduced significantly, but the calculation of the relative contributions of each structure still represents a severe task because of the atomic orbital non-orthogonality. For this reason, a slightly different approach, still chemically motivated, can be made by generalizing the form of the bond functions to include the effects of ionicity. For example, in the case of methane, the bond pair function can be written in the form

$$\varphi_i(1, 2) = \{\lambda[te_i(1)h_i(2) + te_i(2)h_i(1)] + \mu te_i(1)te_i(2)$$
$$+ \xi\, h_i(1)h_i(2)\} \times (1/\sqrt{2})\{\alpha(1)\beta(2) - \beta(1)\alpha(2)\}, \quad (1.20)$$

and the total molecular wave function is still given by the usual anti-symmetrized product of pair functions:

$$\Psi_0'' = N\hat{A}\varphi_1(1, 2)\varphi_2(3, 4)\varphi_3(5, 6)\varphi_4(7, 8).$$

Two of the disadvantages of this formulation are first: the relative weights of multiply ionic structures are fixed by the parameters optimizing the ionicity within a bond. For example, the structure $C^{4-}H_4^+$, appropriate to methane, has a weight $\mu^4$, which may be too large. Secondly, only one spin pairing scheme is considered.

The first limitation is also a characteristic of molecular orbital theory, and does not seem to detract from the usefulness of the results. But in contra-distinction to molecular orbital theory, the model does allow the contributions from ionic structures to tend to zero as the molecule dissociates into atoms.

The practical application of this variant of pair function theory is still hampered by the non-orthogonality of the $\varphi_i$:

$$\iint \varphi_i^*(1, 2)\varphi_j(1, 2)\,\mathrm{d}\tau_1\,\mathrm{d}\tau_2 \neq 0, \quad i \neq j$$

and this has led to attempts to develop valence bond and pair function theories in which all the atomic orbitals have been made mutually orthogonal. However, the use of orthogonal orbitals within valence bond theories requires careful handling, as the following discussion will make clear.

In principle, the non-orthogonality problem can be circumvented by working with the orthogonal orbitals

$$\chi_\lambda = \sum_\alpha A_{\alpha\lambda}\phi_\alpha,$$

where the coefficients $A_{\alpha\lambda}$ are chosen to ensure $\int \chi_\lambda^* \chi_\mu\,\mathrm{d}\tau_1 = \delta_{\lambda\mu}$. The orthogonality between $\chi_\lambda$ and $\chi_\mu$ therefore requires

$$\int \chi^* \chi_\mu\,\mathrm{d}\tau_1 = \sum_\alpha \sum_\gamma A_{\alpha\lambda}^* A_{\gamma\mu} \int \phi_\alpha^* \phi_\gamma\,\mathrm{d}\tau_1 = \delta_{\lambda\mu},$$

i.e.

$$\sum_\alpha \sum_\gamma (\mathbf{A}^+)_{\lambda\alpha} S_{\alpha\gamma} A_{\gamma\mu} = \delta_{\lambda\mu}$$

$$(\mathbf{A}^+\mathbf{S}\mathbf{A})_{\lambda\mu} = \delta_{\lambda\mu}; \quad \lambda, \mu = 1, 2, \ldots$$

yielding

$$\mathbf{A}^+\mathbf{S}\mathbf{A} = \mathbf{1}.$$

The solution of this equation was originally suggested by Löwdin[15] in the form

$$\mathbf{A} = \mathbf{S}^{-1/2}\mathbf{U}$$

where $\mathbf{U}$ is an arbitrary unitary matrix, and $\mathbf{S}^{-1/2}$ is the negative square root of the overlap matrix.† If $\mathbf{U}$ is taken as the unit matrix,[15] then the orthogonal

---

† $\mathbf{S}^{-1/2}$ is determined by diagonalizing $\mathbf{S}$, taking the inverse square root of each diagonal element, and then performing the reverse transformation with the unitary matrix which diagonalizes $\mathbf{S}$: If $\mathbf{s} = \mathbf{U}^+\mathbf{S}\mathbf{U}$, where $(\mathbf{s})_{\alpha\gamma} = \lambda_\alpha\delta_{\alpha\gamma}$, then $(\mathbf{s}^{-1/2})_{\alpha\gamma} = \lambda^{-1/2}\delta_{\alpha\gamma}$, and hence $\mathbf{S}^{-1/2} = \mathbf{U}\mathbf{s}^{-1/2}\mathbf{U}^+$.

functions are obtained in the form

$$\chi_\lambda = \sum_\alpha (S^{-1/2})_{\alpha\lambda}\phi_\alpha. \tag{1.21}$$

The use of the orthogonal functions, $\chi_\lambda$, leads to a simplification of the energy calculation, but much of the binding energy is lost if the ground state wave function is approximated by a simple Heitler–London type of covalent structure. The reasons for this failing of the theory have been analysed in detail by McWeeny,[16] but the problem was first noted by Slater[17] in a calculation on the hydrogen molecule with a Heitler–London-like wave function of the form

$$\Psi_0 = (1/\sqrt{2})\{\chi_1(1)\chi_2(2) + \chi_1(2)\chi_2(1)\} \times (1/\sqrt{2})\{\alpha(1)\beta(2) - \beta(1)\alpha(2)\}. \tag{1.22}$$

If $(S^{-1/2})_{11}$ and $(S^{-1/2})_{22}$ are denoted by $x$, and $(S^{-1/2})_{12}$, $(S^{-1/2})_{21}$ by $y$, then substitution of (1.21) into (1.22) shows that

$$\Psi_0 = \{(x^2 + y^2)[\phi_1(1)\phi_2(2) + \phi_1(2)\phi_2(1)]$$
$$+ 2xy[\phi_1(1)\phi_1(2) + \phi_2(1)\phi_2(2)]\} \cdot (1/2)\{\alpha(1)\beta(2) - \beta(1)\alpha(2)\}$$

The weighting of the polar structures is therefore fixed by the value of the overlap integral between $\phi_1$ and $\phi_2$. The correct balance between covalent and ionic structures is obtained only after explicit inclusion of ionic structures involving $\chi_1$ and $\chi_2$, with the consequent determination of the weighting coefficients by energy minimization. In fact, calculations based on the use of either $\phi_1$, $\phi_2$ or $\chi_1$, $\chi_2$ give the same results when all structures are considered. But although the simple Heitler–London structure, based on the use of $\phi_1$, $\phi_2$, gives fairly reasonable results for the energy as a function of the internuclear separation, very poor results are obtained when the symmetrically orthogonalized functions, $\chi_\lambda$, are used. The reason for this behaviour is that the Heitler–London-like structure involving $\chi_1$, $\chi_2$ does not result in a build-up of charge density in between the two nuclei: a prerequisite for the formation of a strong bond (see Vol. 1). Thus the simplifications introduced by the use of orthogonal functions are lost to a large extent by the necessity of working with an extensive set of structures.

Although the non-orthogonality problem still appears impassable, with applications apparently limited to diatomic molecules, the framework of the theory—particularly within the formalism of a pair function approach—is of considerable chemical interest, as the wave function portrays in a direct manner the intuitive notions of bonds and lone-pairs. In fact, there have been attempts, particularly by McWeeny and Klessinger,[18] to evolve a practical form of pair wave function having these characteristics.

The usual assumption is made whereby the non-valence electrons are treated as a non-polarizable core. The valence atomic orbitals, centred on the different nuclei, are then Schmidt orthogonalized to all core atomic orbitals (this orthogonalization procedure is a generalization of the one used in constructing the orthogonal spin functions for the four spin system; see also Vol. 1). In the case of $BH_2^+$, for example, where $h_1$ and $h_2$ are not orthogonal to $1s_B$, the wave function is unaltered if $h_i$ is replaced by $\chi_i = ah_i - b1s_B$, where $a$ and $b$ are chosen so that $\chi_i$ is normalized and orthogonal to $1s_B$. This result is best verified by expanding the pair wave function in terms of determinants, and noticing that the introduction of the $\chi_i$ merely involves replacing rows in each determinant by linear combinations of rows—a manipulation which does not alter the value of a determinant.

The orthogonalization of the valence atomic orbitals to the core atomic orbitals is crucial if the core electrons are not included explicitly in the molecular wave function. The calculations of Stuart and Hurst[19] on LiH show very clearly that large errors are introduced if the molecule is treated as a four-electron problem, without orthogonalizing the hydrogen $1s$ function to the lithium $1s$ core function.

The Löwdin[15] procedure is now used to construct a set of orthogonal hybrid orbitals, from the set of valence atomic orbitals which have been orthogonalized to all core atomic orbitals. The resulting orbitals are then used to construct pair functions of the form (1.20). However, it must be remembered that these pair functions only describe bonds and lone-pairs in a formal sense, as the orthogonalized atomic orbitals are built out of the ordinary atomic orbitals centred on the various nuclei in the molecule. The inclusion of ionic structures, within each bond function, is therefore of the utmost importance for obtaining an adequate description of the electron distribution, as already discussed in the case of the hydrogen molecule. Some of the results obtained by Klessinger[20-21] and by Franchini and Vergani[22] are given in Table 1.1. The calculated molecular energies are always better than those obtained from molecular orbital calculations.

TABLE 1.1

*Comparison of molecular properties as calculated from strongly orthogonal pair wave functions (SOPF) and single configuration molecular orbital wave functions (SCMO)*

|  |  | HF | $H_2O$ | $NH_3$ | $CH_4$ |
|---|---|---|---|---|---|
| Total energy, $\epsilon$ | SOPF | $-99\cdot4972$ | $-75\cdot6902$ | $-56\cdot0511$ | $-40\cdot1709$ |
| (a.u.) | SCMO | $-99\cdot4786$ | $-75\cdot6617$ | $-56\cdot0045$ | $-40\cdot1129$ |
| Dipole moment | SOPF | $0\cdot808$ | $1\cdot612$ | $1\cdot899$ | — |
| (Debyes) | SCMO | — | $1\cdot630$ | $1\cdot927$ | — |
|  | Expt. | $1\cdot82$ | $1\cdot85$ | $1\cdot47$ | — |
| $R_{HX} = \lvert \mathbf{R}_H - \mathbf{R}_X \rvert$ |  | $0\cdot0917$ | $0\cdot0958$ | $0\cdot101$ | $0\cdot109$ |
| (nm) |  |  |  |  |  |
| $H\hat{X}H$ |  | — | $105°$ | $107°$ | $109° \, 27'$ |
| References |  | 21, 54 | 20 | 22 | 22 |

In its present form, the theory cannot be applied to molecules where electron delocalization is known, or thought, to be important, as interbond charge transfer structures are excluded.

Since the pair functions are orthogonal in both the strong and the weak senses

$$\int \varphi_i(1, 2)\varphi_j(1, 3)\, d\tau_1 = 0; \quad \int \varphi_i(1, 2)\varphi_j(1, 2)\, d\tau_1\, d\tau_2 = 0, \quad i \neq j$$

the expression for the molecular energy reduces to a particularly simple form. For on writing $\Psi_0$ as

$$N\hat{A}\varphi_1(1, 2)\varphi_2(3, 4)\ldots\varphi_m(2m - 1, 2m)$$
$$= N2^{-m/2}\hat{A}(1 - \hat{P}_{12})(1 - \hat{P}_{34})\ldots(1 - \hat{P}_{2m-12m})\psi_1(1, 2)\ldots$$
$$\times \psi_m(2m - 1, 2m) \times \alpha(1)\beta(2)\ldots\alpha(2m - 1)\beta(2m)$$
$$= N2^{m/2}\hat{A}\psi_1(1, 2)\ldots\psi_m(2m - 1, 2m)\alpha(1)\ldots\beta(2m), \tag{1.23}$$

where

$$\varphi_i(1, 2) = \psi_i(1, 2).(1/\sqrt{2})[\alpha(1)\beta(2) - \beta(1)\alpha(2)]$$
$$= 2^{-1/2}(1 - \hat{P}_{12})\psi_i(1, 2)\alpha(1)\beta(2),$$

the energy expression becomes

$$E = \frac{\begin{aligned}\int \hat{A}\psi_1(1, 2)\ldots\psi_m(2m - 1, 2m)\alpha(1)\beta(2)\ldots\alpha(2m - 1)\beta(2m) \times \\ \times \hat{H}\hat{A}\psi_1(1, 2)\ldots\psi_m(2m - 1, 2m) \times \\ \times \alpha(1)\beta(2)\ldots\alpha(2m - 1)\beta(2m)\, d\tau\end{aligned}}{\begin{aligned}\int \hat{A}\psi_1(1, 2)\ldots\psi_m(2m - 1, 2m)\alpha(1)\beta(2)\ldots\alpha(2m - 1)\beta(2m) \times \\ \times \hat{A}\psi_1(1, 2)\ldots\psi_m(2m - 1, 2m) \times \\ \times \alpha(1)\beta(2)\ldots\alpha(2m - 1)\beta(2m)\, d\tau\end{aligned}} \tag{1.24}$$

First, consider the overlap integral

$$\int \left(\sum_{\hat{P}}(-1)^P\hat{P}\psi_1\ldots\psi_m\alpha(1)\ldots\beta(2m)\right)\left(\sum_{\hat{Q}}(-1)^Q\hat{Q}\psi_1\ldots\psi_m\alpha(1)\ldots\beta(2m)\right)d\tau$$

forming the denominator of (1.24). Non-zero contributions to the integral are obtained only if the permutation $\hat{P}$ matches the permutation $\hat{Q}$: for if $\hat{P}$ and $\hat{Q}$ are different, then the corresponding contribution to the integral is zero because of the orthogonality of either the spin or the orbital functions. Now any one of the $(2m)!$ permutations, $\hat{P}$, can be matched with the corresponding permutations, $\hat{Q}$, and since each term, on integration, yields unity, the denominator of (1.24) has the value

$$(2m)!.$$

The integrations in the numerator of the energy expression are best performed by rewriting the molecular Hamiltonian in the form (see (1.3))

$$\hat{H} = \sum_i \hat{f}_i + \sum_{i<j} \frac{1}{r_{ij}} = \sum_{i<j} \left[ \frac{1}{2m-1}\hat{f}_i + \frac{1}{2m-1}\hat{f}_j + \frac{1}{r_{ij}} \right] = \sum_{i<j} \hat{h}_{ij}$$

where $\hat{f}_i$ is the one-electron operator

$$-\tfrac{1}{2}\nabla_i^2 - \sum_\alpha \frac{Z_\alpha}{r_{\alpha i}}$$

Consider the term in $\hat{H}$ arising from the pair of electrons labelled 1 and 2:

$$\int \left( \sum_{\hat{P}} (-1)^P \hat{P} \psi_1 \ldots \psi_m \alpha(1) \ldots \beta(2m) \right) \hat{h}_{12} \times$$

$$\times \left( \sum_{\hat{Q}} (-1)^Q \hat{Q} \psi_1 \ldots \psi_m \alpha(1) \ldots \beta(2m) \right) d\tau.$$

In the evaluation of this integral, only those permutations $\hat{P}$ and $\hat{Q}$ need be considered which differ in the allocation of electrons to at most two pair functions—any mismatches of higher order yield zero after integration because of the orthogonality requirements. For a given permutation, $\hat{P}$, this restricts $\hat{Q}$ to a permutation involving the redistribution of the electron labels within two pair functions: for example, consider the situation in which electrons labelled 1 and 2 are placed in the pair functions $\psi_1$ and $\psi_2$, and let the remaining two electrons in these pair functions be labelled by $i$ and $j$. Out of the $(4!)^2$ possible integrals involving different allocations of 1, 2, $i$ and $j$ to $\psi_1$ and $\psi_2$, only a small number survive because of the orthogonality of either the orbital or the spin functions. This means that unless the electron labels $i$ and $j$ occupy the same relative positions in identical pair functions, there is no contribution to the energy. For example,

$$\int \psi_1(1,i)\psi_2(2,j)\hat{h}_{12}\psi_1(1,j)\psi_2(2,i)\alpha(1)\beta(i)\alpha(2)\beta(j)\alpha(1)\beta(j)\alpha(2)\beta(i) \times$$

$$\times d\tau_1\, d\tau_2\, d\tau_i\, d\tau_j$$

gives zero after integrating over the spatial variables of electrons labelled $i$ and $j$.

The 4! matching permutations give rise to the four terms

$$\int \psi_1(1,2)\psi_2(i,j)\hat{h}_{12}\psi_1(1,2)\psi_2(i,j)\alpha(1)\beta(2)\alpha(i)\beta(j)\alpha(1)\beta(2)\alpha(i)\beta(j) \times$$

$$\times d\tau_1\, d\tau_2\, d\tau_i\, d\tau_j,$$

$$\int \psi_1(2,1)\psi_2(i,j)\hat{h}_{12}\psi_1(2,1)\psi_2(i,j)\alpha(2)\beta(1)\alpha(i)\beta(j)\alpha(2)\beta(1)\alpha(i)\beta(j) \times$$

$$\times d\tau_1\, d\tau_2\, d\tau_i\, d\tau_j,$$

$$\int \psi_1(2,1)\psi_2(j,i)\hat{h}_{12}\psi_1(2,1)\psi_2(j,i)\alpha(2)\beta(1)\alpha(j)\beta(i)\alpha(2)\beta(1)\alpha(j)\beta(i) \times$$

$$\times d\tau_1\, d\tau_2\, d\tau_i\, d\tau_j,$$

$$\int \psi_1(1,2)\psi_2(j,i)\hat{h}_{12}\psi_1(1,2)\psi_2(j,i)\alpha(1)\beta(2)\alpha(j)\beta(i)\alpha(1)\beta(2)\alpha(j)\beta(i) \times$$

$$\times d\tau_1\, d\tau_2\, d\tau_i\, d\tau_j,$$

where the electron labels 1 and 2 are associated with the pair function $\psi_1$. There are another four terms in which the electron labels 1 and 2 are associated with $\psi_2$. The remaining sixteen permutations all give rise to integrals of the form

$$\int \psi_1(1, i)\psi_2(2, j)\hat{h}_{12}\psi_1(1, i)\psi_2(2, j) \, d\mathbf{r}_1 \, d\mathbf{r}_2 \, d\mathbf{r}_i \, d\mathbf{r}_j$$

after integrating over spin coordinates.

Finally, there are some additional contributions coming from mismatching permutations of 1, 2, $i$ and $j$ which differ only in the allocation of electrons with the same spin:

$$- 8 \int \psi_1(1, i)\psi_2(2, j)\hat{h}_{12}\psi_1(2, i)\psi_2(1, j) \, d\mathbf{r}_1 \, d\mathbf{r}_2 \, d\mathbf{r}_i \, d\mathbf{r}_j.$$

The contribution of the above permutations to that part of the molecular energy involving $\hat{h}_{12}$ can be written in the form

$$\frac{1}{(2m)!} \left\{ 4 \int \psi_1(1, 2)\hat{h}_{12}\psi_1(1, 2) \, d\mathbf{r}_1 \, d\mathbf{r}_2 \right.$$

$$+ 4 \int \psi_2(1, 2)\hat{h}_{12}\psi_2(1, 2) \, d\mathbf{r}_1 \, d\mathbf{r}_2$$

$$+ 4 \int \rho_1(1, 1)\hat{h}_{12}\rho_2(2, 2) \, d\mathbf{r}_1 \, d\mathbf{r}_2$$

$$\left. - 2 \int \rho_1(1, 2)\hat{h}_{12}\rho_2(1, 2) \, d\mathbf{r}_1 \, d\mathbf{r}_2 \right\} \qquad (1.25)$$

where some of the integrals have been simplified in appearance by introducing the density functions

$$\rho_i(1, 2) = 2 \int \psi_i(1, 3)\psi_i(2, 3) \, d\mathbf{r}_3. \quad (i = 1, 2)$$

Now each matching permutation of $1, 2, i$ and $j$ can be achieved in $(2m - 4)!$ ways, as the sequence of the remaining $(2m - 4)$ electron labels is of no interest, unless the two sequences produced by $\hat{P}$ and $\hat{Q}$ are matching. In addition, the arbitrary electron labels $i$ and $j$ can be selected in $^{2m-2}C_2$ ways. Therefore (1.25) requires multiplying by $(2m - 4)!\,^{2m-2}C_2$ to give the contribution to the energy arising from the allocation of the electrons labelled 1 and 2 to $\varphi_1$ and $\varphi_2$. The total contribution to the molecular energy arising from $\hat{h}_{12}$ is found by allowing the electrons labelled 1 and 2 to occupy any two pair functions:

$$\frac{2}{2m(2m - 1)} \left\{ \sum_i \int \psi_i(1, 2)\hat{h}_{12}\psi_i(1, 2) \, d\mathbf{r}_1 \, d\mathbf{r}_2 + \sum_{i<j} \int \rho_i(1, 1)\hat{h}_{12} \times \right.$$

$$\left. \times \rho_j(2, 2) \, d\mathbf{r}_1 \, d\mathbf{r}_2 - \tfrac{1}{2} \sum_{i<j} \int \rho_i(1, 2)\hat{h}_{12}\rho_j(1, 2) \, d\mathbf{r}_1 \, d\mathbf{r}_2 \right\}. \quad (1.26)$$

The same result is also obtained for any other choice of $\hat{h}_{ij}$, except that the names of the integration variables are changed to $i$ and $j$, rather than 1 and 2; and this cannot affect the values of the integrals. Since there are $2m(2m - 1)/2$

ways of selecting a pair of electrons, (1.26) requires multiplying by this number in order to obtain the total electronic energy:

$$
\begin{aligned}
E = &\sum_i \int \psi_i(1, 2) \left[ \frac{\hat{f}_1}{2m-1} + \frac{\hat{f}_2}{2m-1} + \frac{1}{r_{12}} \right] \psi_i(1, 2) \, \mathrm{d}r_1 \, \mathrm{d}r_2 \\
&+ \sum_{i<j} \int \rho_i(1, 1) \left[ \frac{\hat{f}_1}{2m-1} + \frac{\hat{f}_2}{2m-1} \right] \rho_j(2, 2) \, \mathrm{d}r_1 \, \mathrm{d}r_2 \\
&- \frac{1}{2} \sum_{i<j} \int \rho_i(1, 2) \frac{1}{r_{12}} \rho_j(1, 2) \, \mathrm{d}r_1 \, \mathrm{d}r_2 \\
&\qquad\qquad + \sum_{i<j} \int \rho_i(1, 1) \frac{1}{r_{12}} \rho_j(2, 2) \, \mathrm{d}r_1 \, \mathrm{d}r_2 \\
= &\sum_i \int \psi_i(1, 2) \left[ \hat{f}_1 + \hat{f}_2 + \frac{1}{r_{12}} \right] \psi_i(1, 2) \, \mathrm{d}r_1 \, \mathrm{d}r_2 \\
&+ \sum_{i<j} \int \rho_i(1, 1) \frac{1}{r_{12}} \cdot \rho_j(2, 2) \, \mathrm{d}r_1 \, \mathrm{d}r_2 \\
&\qquad\qquad - \frac{1}{2} \sum_{i<j} \int \rho_i(1, 2) \frac{1}{r_{12}} \cdot \rho_j(1, 2) \, \mathrm{d}r_1 \, \mathrm{d}r_2 \quad (1.27)
\end{aligned}
$$

where use is made of the relation $\int \rho_i(2, 2) \, \mathrm{d}r_2 = 2$.

Equation (1.27) shows clearly how the total energy is partitioned into intrapair and interpair contributions. But, as discussed previously, this is only a formal partition, as the basic one-electron orthogonal orbitals are built out of atomic orbitals centred on the various nuclei in the molecule.

An alternative expression to (1.27) has been used for many years, for the special situation in which ionic contributions to each bond pair function are neglected. If the molecule contains $M$ core-pairs, $P$ lone-pairs, and $N$ bond-pairs, then the expression for the molecular energy can be rewritten in the form

$$
\begin{aligned}
E = &\sum_{i=1}^{M} \int \psi_i \hat{h}_{12} \psi_i \, \mathrm{d}r + \sum_{i=M+1}^{M+P} \int \psi_i \hat{h}_{12} \psi_i \, \mathrm{d}r + \sum_{i=M+P+1}^{M+P+N} \int \psi_i \hat{h}_{12} \psi_i \, \mathrm{d}r \\
&+ \sum_{i<j=M+1}^{M+P} X_{ij} + \sum_{i<j=M+P+1}^{M+P+N} X_{ij} + \sum_{i=M+1}^{M+P} \sum_{j=M+P+1}^{M+P+N} X_{ij} \\
&+ \sum_{i=1}^{M} \left\{ \sum_{j=M+1}^{M+P} X_{ij} + \sum_{j=M+P+1}^{M+P+N} X_{ij} \right\} + \sum_{i<j=1}^{M} X_{ij}
\end{aligned}
$$

where the $X_{ij}$ are defined by

$$
X_{ij} = \int \rho_i(1, 1) \frac{1}{r_{12}} \rho_j(2, 2) \, \mathrm{d}r - \tfrac{1}{2} \int \rho_i(1, 2) \frac{1}{r_{12}} \rho_j(1, 2) \, \mathrm{d}r
$$

and the various $\rho_i$ are given by

$$
\rho_C(1, 2) = 2\chi_C(1)\chi_C(2); \quad \rho_l(1, 2) = 2\chi_L(1)\chi_L(2);
$$
$$
\rho_b(1, 2) = \chi_B(1)\chi_B(2) + \chi_{B''}(1)\chi_{B''}(2).
$$

The suffixes $c$, $l$, and $b$ label typical core, lone-pair, and bonding functions, respectively; and the suffices $C$, $L$, and $B$ label the basic atomic orbitals according to

$$c = 1, 2, \ldots M,$$
$$l = M + 1, M + 2, \ldots M + P,$$
$$b = M + P + 1, \ldots P + M + N,$$

$$C = 1, 2, \ldots M$$
$$L = M + 1, \ldots M + P$$
$$B = M + P + 1, \ldots M + 2N + P$$

(two atomic orbitals are required to define each bond pair). The labels $B$ and $B''$ refer to the two atomic orbitals involved in the bond pair function $\varphi_B$.

The molecular energy is now given by

$$E = \sum_B \int \chi_B \hat{f} \chi_B \, d\mathbf{r}_1 + \sum_{\substack{B, B'' \\ \text{(pairs)}}} (J_{BB''} + K_{BB''}) + 2 \sum_L \int \chi_L \hat{f} \chi_L \, d\mathbf{r}_1$$

$$+ \sum_L J_{LL} + 2 \sum_C \int \chi_C \chi \hat{f}_C \, d\mathbf{r}_1 + \sum_C J_{CC}$$

$$+ \sum_{C, B} (2J_{CB} - K_{CB}) + \sum_{C, L} (4J_{CL} - 2K_{CL})$$

$$+ \sum_{L, B} (2J_{LB} - K_{LB}) + \sum_{\substack{\text{pairs } B, B' \\ (B' \neq B'')}} (J_{BB'} - \tfrac{1}{2}K_{BB'})$$

$$+ \tfrac{1}{2} {\sum_{C, C'}}' (4J_{CC'} - 2K_{CC'}) + \tfrac{1}{2} {\sum_{L, L'}}' (4J_{LL'} - 2K_{LL'}) \quad (1.28)$$

where $J_{XY}$ and $K_{XY}$ denote the Coulomb and exchange integrals

$$\int \chi_X(1)\chi_Y(2)(1/r_{12})\chi_X(1)\chi_Y(2) \, d\mathbf{r}_1 \, d\mathbf{r}_2$$

and

$$\int \chi_X(1)\chi_Y(2)(1/r_{12})\chi_Y(1)\chi_X(2) \, d\mathbf{r}_1 \, d\mathbf{r}_2,$$

respectively.

Equation (1.28) can be written in a more compact form by introducing the orbital occupation numbers, $n_X$ $(X = C, L, B)$, and the notation $f_{XX} = \int \chi_X \hat{f} \chi_X \, d\mathbf{r}_1$:

$$E = \sum_X n_X f_{XX} + \sum_{\substack{\text{pairs} \\ X, Y}} n_X n_Y (J_{XY} - \tfrac{1}{2}K_{XY}) + \tfrac{3}{2} \sum_{\substack{\text{pairs} \\ B, B''}} K_{BB''}$$

$$+ \tfrac{1}{2} \sum_{\substack{X \\ (\neq B)}} n_X J_{XX}, \quad (1.29)$$

where it should be noted that the summation over the pairs $X$, $Y$ includes the terms with $X = B$ and $Y = B''$. This equation is formally similar to equation (25) given by Coulson[23] which, as he describes, was originally derived by Dirac.[24] The equation has been used widely—particularly in stereochemical discussions—and the reader is referred to Coulson[23] for specific examples, and further details. The equation also forms the basis for the discussion of valence state energies which is given in the next chapter.

It should be remembered, however, that the total energy is poorly represented by a Heitler–London type of wave function which is based on the use of orthogonal atomic orbitals. This difficulty is usually circumvented by using empirical values for the integrals in (1.29), and all deficiencies of the model are then hopefully absorbed through the parameterization procedure. But this is basically unsatisfactory, and really precludes any quantitative discussion of the molecular electronic structure. For this reason, a more complete pair theory is favoured which includes explicit contributions from intra-bond ionic structures. However, at the moment, applications of the model are limited to the ground states of small polyatomic molecules, and it remains to be seen whether the method can be used successfully for larger molecules, and for discussing excited states.

The discussion of valence bond based theories is now concluded, and the remainder of the chapter is concerned with molecular orbital theory and the methods used for describing molecular electronic charge distributions.

## 1.5.   The method of molecular orbitals

In the molecular orbital theory, as applied to a closed shell molecule, the approximate ground-state wave function is written as an antisymmetrized product of doubly occupied one-electron orbitals, $\psi_i$, which are, in general, delocalized over the whole molecular framework.

$$\Psi_0 = N\hat{A}\psi_1(1)\bar{\psi}_1(2) \ldots \psi_m(2m - 1)\bar{\psi}_m(2m). \qquad (1.30)$$

As before, and in Vol. 1, the presence or absence of a bar sign implies a spin function with $m_s = -\frac{1}{2}$ or $m_s = +\frac{1}{2}$, respectively.

Thus, at this level of approximation, each electron is assumed to move independently in the potential field of the nuclei, and the average potential of the remaining electrons. Consequently, electrons with opposed spins are, on average, not kept far enough apart from each other: this lack of correlation is particularly serious for pairs of electrons in the same molecular orbital. On the other hand, electrons with parallel spins are partially correlated because of the antisymmetric form of the molecular wave function (see Vol. 1).

The inadequate treatment of electron correlation in molecular orbital theory should be contrasted with the high degree of correlation present in pair theories, based solely on the use of Heitler–London bond functions: for here, each electron is on average associated with a different nucleus. The inclusion of ionic structures, within the valence bond approach, represents an attempt to improve the degree of electron correlation. But, as discussed in Vol. 1, this is in direct contradistinction to molecular orbital theory, where the weighting of ionic structures is fixed by the form of the molecular orbitals, and is not determined variationally.

The poor treatment of electron correlation, arising from the use of the molecular orbital wave function (1.33), is a direct consequence of requiring pairs of electrons to occupy the same molecular orbital. This requirement can be relaxed, though, by allowing electrons with different components of spin angular momentum to occupy different spatial orbitals. But a further difficulty now arises, because the single determinantal wave function

$$\Psi_0 = N\hat{A}\psi_i(1)\bar{\psi}_1'(2) \ldots \psi_m(2m-1)\bar{\psi}_m'(2m)$$

although an eigenfunction of $\hat{S}_z$ is no longer an eigenfunction of $\hat{S}^2$. In principle, it is therefore necessary to construct an appropriate singlet function before optimizing the molecular energy. A similar situation obtains for molecules with open shells, where the ground electronic state cannot be represented by a single determinant. There are severe problems in handling molecular wave functions in the "open shell" form—some of which are still incompletely resolved—and so most of this chapter is restricted to a discussion of the electronic structure of molecules whose ground state wave functions may be approximated in the manner of (1.33). In this approach, the one-electron orbitals, $\psi_i$, are delocalized over the molecular framework, and any notion of bonds is sacrificed in preference for mathematical simplicity. However, localized *one-electron* functions can be obtained from the set of delocalized one-electron functions by a suitable unitary transformation.[25]

The optimum form for each $\psi_i$ is determined by applying the variation theorem to the expression for the total electronic energy (see Appendix I, after allowing for the use of doubly occupied orthogonal orbitals):

$$E = 2\sum_i \int \psi_i(1)\hat{f}_1\psi_i(1)\,\mathrm{d}r_1 + \sum_{i,j}(2J_{ij} - K_{ij}). \tag{1.31}$$

$J_{ij}$ and $K_{ij}$ are the usual Coulomb and exchange integrals as defined in the previous section, except that the symbol $\psi$ now replaces $\chi$. Equation (1.31) can be written in a more compact form if Coulomb and exchange operators, $\hat{J}_j$ and $\hat{K}_j$ respectively, are first introduced by the equations

$$\iint \psi_i(1)\psi_j(2)(1/r_{12})\psi_i(1)\psi_j(2)\,\mathrm{d}r_1\,\mathrm{d}r_2 = \int \psi_i(1)\hat{J}_j(1)\psi_i(1)\,\mathrm{d}r_1$$

$$\iint \psi_i(1)\psi_j(2)(1/r_{12})\psi_j(1)\psi_i(2)\,\mathrm{d}r_1\,\mathrm{d}r_2 = \int \psi_i(1)\hat{K}_j(1)\psi_i(1)\,\mathrm{d}r_1.$$

Notice that $\hat{K}_j$ is an integral operator since

$$\hat{K}_j(1)\psi_i(1) = \int \frac{\psi_j(2)\psi_j(1)}{r_{12}}\,\psi_i(2)\,\mathrm{d}r_2,$$

i.e.

$$\hat{K}_j(1) \equiv \int \frac{\psi_j(2)\hat{P}_{12}\psi_j(2)}{r_{12}}\,\mathrm{d}r_2$$

where $\hat{P}_{12}$ is the operator which interchanges electrons 1 and 2. Thus, (1.31) becomes

$$E = 2 \sum_i \int \psi_i \left[ \hat{f} + \sum_j (\hat{J}_j - \tfrac{1}{2}\hat{K}_j) \right] \psi_i \, dr_1$$

$$= 2 \sum_i \int \psi_i \hat{h}_a \psi_i \, dr_1 \qquad (1.32)$$

where the electron labels have been dropped for simplicity. Equation (1.32) describes an additive partitioning of the electronic energy.

The application of the variation theorem leads to

$$\delta E = 2 \sum_i \int \delta\psi_i \hat{h}_a \psi_i \, dr_1 + 2 \sum_i \int \psi_i \hat{h}_a \delta\psi_i \, dr_1 + 2 \sum_i \int \psi_i \delta\hat{h}_a \psi_i \, dr_1 \quad (1.33)$$

remembering that $\hat{h}_a$ depends upon the unknown $\psi_i$ (assumed real).

On expanding the last term in (1.33) in terms of the basic two-electron integrals, and then reintroducing the Coulomb and exchange operators, the expression for $\delta E$ becomes

$$\delta E = 4 \sum_i \int \delta\psi_i \hat{h}_a \psi_i \, dr_1 + 4 \sum_i \int \delta\psi_i \left( \sum_j (\hat{J}_j - \tfrac{1}{2}\hat{K}_j) \right) \psi_i \, dr_1$$

which, after introducing the effective one-electron operator $\hat{h}$ by

$$\hat{h} = \hat{h}_a + \sum_j (\hat{J}_j - \tfrac{1}{2}\hat{K}_j) = \hat{f} + 2 \sum_j (\hat{J}_j - \tfrac{1}{2}\hat{K}_j),$$

reduces to

$$\delta E = 4 \sum_i \int \delta\psi_i \hat{h} \psi_i \, dr_1.$$

However, the $\psi_i$ must remain orthonormal during the energy variation, as any deviation from orthonormality invalidates the use of (1.32) for calculating the new energy with the orbitals $\psi_i + \delta\psi_i$. The orthonormality is maintained by using one Lagrangian multiplier, $(- 2\epsilon_{ij})$, for each constraint $\int \psi_i\psi_j \, dr_1 = \delta_{ij}$. The permitted first-order variation in the energy, consistent with the retention of orthonormality, is therefore given by

$$\delta E = 4 \sum_i \int \delta\psi_i \hat{h}\psi_i \, dr_1 - 4 \sum_{j,i} [(\epsilon'_{ij} + \epsilon'_{ji})/2] \int \delta\psi_i\psi_j \, dr_1$$

and this must vanish at the energy minimum: that is,

$$\sum_i \int \delta\psi_i \left[ \hat{h}\psi_i - \sum_j \epsilon_{ji}\psi_j \right] dr_1 = 0$$

where $(\epsilon'_{ij} + \epsilon'_{ji})/2 = \epsilon_{ji}$.

As the $\delta\psi_i$ represent arbitrary independent variations in the $\psi_i$, each of the terms in the summation over the index $i$ can vanish only if the $\psi_i$ satisfy the coupled equations

$$\hat{h}\psi_i = \sum_j \epsilon_{ji}\psi_j \qquad i = 1, 2, \ldots, m. \qquad (1.34)$$

where, for real $\psi_i$, the $\epsilon_{ji}$ are elements of a symmetric matrix (see above).

Now the total molecular wave function is determined only to within a phase factor of modulus unity: for if the orthogonal transformation

$$\psi_i \to \psi_i^{\mathrm{or}} = \sum_j O_{ji}\psi_j \tag{1.35}$$

is made on the occupied $\psi_i$, then

$$\Psi_0 \to \Psi_0^{\mathrm{or}} = \left(\sum_{j,\,k,\,l\ldots} O_{j1}O_{k2}O_{l3}\ldots\right)\left(\sum_{p,\,q,\,r\ldots} O_{p1}O_{q2}O_{r3}\ldots\right)\times$$
$$\times\, N\hat{A}\psi_j(1)\bar{\psi}_p(2)\psi_k(3)\bar{\psi}_q(4)\ldots$$
$$= \left(\sum_{j,\,k,\,l\ldots} (-1)^P O_{j1}O_{k2}O_{l3}\ldots\right)\left(\sum_{p,\,q,\,r\ldots} (-1)^Q O_{p1}O_{q2}O_{r3}\ldots\right)\times$$
$$\times\, N\hat{A}\psi_1(1)\bar{\psi}_1(2)\psi_2(3)\bar{\psi}_2(4)\ldots$$

where $(-1)^P$ and $(-1)^Q$ are the parities of the permutations required to produce the sequence $\psi_1\psi_2\psi_3\ldots$ from $\psi_j\psi_k\psi_l\ldots$ and $\psi_p\psi_q\psi_r\ldots$, respectively. Hence, $\Psi_0^{\mathrm{or}}$ becomes

$$\Psi_0^{\mathrm{or}} = (\det \mathbf{O})^2 N\hat{A}\psi_1(1)\bar{\psi}_1(2)\psi_2(3)\bar{\psi}_2(4)\ldots = \Psi_0.$$

The total energy also remains invariant under the transformation (1.35):

$$E \to E^{\mathrm{or}} = 2\sum_i \int \psi_i^{\mathrm{or}}\left[\hat{f} + \sum_j (\hat{J}_j^{\mathrm{or}} - \tfrac{1}{2}\hat{K}_j^{\mathrm{or}})\right]\psi_i^{\mathrm{or}}\,d\mathbf{r}_1,$$

and since

$$\sum_j (\hat{J}_j^{\mathrm{or}} - \tfrac{1}{2}\hat{K}_j^{\mathrm{or}}) = \sum_{j,\,p,\,q} O_{pj}O_{qj}\left[\int \frac{\psi_p(1)[1 - \tfrac{1}{2}\hat{P}_{12}]\psi_q(1)}{r_{12}}\,d\mathbf{r}_1\right]$$
$$= \sum_{p,\,q} \delta_{pq}\int \psi_p(1)\frac{1}{r_{12}}[1 - \tfrac{1}{2}\hat{P}_{12}]\psi_q(1)\,d\mathbf{r}_1$$
$$= \sum_p (\hat{J}_p - \tfrac{1}{2}\hat{K}_p)$$

$\hat{h}_a^{\mathrm{or}}$ reduces to $\hat{h}_a$, resulting in

$$E^{\mathrm{or}} = 2\sum_{i,\,s,\,t} O_{si}O_{ti}\int \psi_s\hat{h}_a\psi_t\,d\mathbf{r}_1 = 2\sum_s \int \psi_s\hat{h}_a\psi_s\,d\mathbf{r}_1 = E.$$

It is now apparent that the orbitals $\psi_i^{\mathrm{or}}$ also satisfy (1.34), since

$$\hat{h}^{\mathrm{or}}\psi_i^{\mathrm{or}} = \hat{h}\psi_i^{\mathrm{or}} = \sum_j \epsilon_{ji}^{\mathrm{or}}\psi_j^{\mathrm{or}} \qquad (i = 1, 2, \ldots m). \tag{1.36}$$

This result is important because the degree of freedom, expressed by the transformation (1.35), can be used to uncouple the equations (1.34). But before proving this result, it is convenient to rewrite the equations (1.36) in the more compact matrix form

$$\hat{h}\boldsymbol{\psi}^{\mathrm{or}} = \boldsymbol{\psi}^{\mathrm{or}}\boldsymbol{\epsilon}^{\mathrm{or}} \tag{1.37}$$

where the $\psi_i^{\mathrm{or}}$ are collected into the row matrix $\boldsymbol{\psi}^{\mathrm{or}}$. In addition, the transformation between the $\psi_i$ and the $\psi_i^{\text{or}}$ is given by $\boldsymbol{\psi}^{\text{or}} = \boldsymbol{\psi O}$. Thus, substitution for $\boldsymbol{\psi}^{\text{or}}$ in (1.37), and post-multiplication by $\mathbf{O}'$ yields

$$\hat{h}\boldsymbol{\psi} = \boldsymbol{\psi O}\,\boldsymbol{\epsilon}^{\text{or}}\mathbf{O}' = \boldsymbol{\psi\epsilon}$$

where the last step follows from the matrix form of (1.34). The matrix $\boldsymbol{\epsilon}^{\text{or}}$ is therefore related to $\boldsymbol{\epsilon}$ by means of the transformation $\boldsymbol{\epsilon}^{\text{or}} = \mathbf{O}'\boldsymbol{\epsilon}\mathbf{O}$. The orthogonal matrix, $\mathbf{O}$, which diagonalizes $\boldsymbol{\epsilon}$ defines the one-electron functions, $\psi_i^{\text{or}}$, termed molecular orbitals.† A suitable transformation always exists, as any symmetric matrix can be reduced to diagonal form by some orthogonal transformation.[10] The original equations (1.34) are therefore only uncoupled for one particular choice of $\mathbf{O}$. Alternatively, if equations are assumed to be uncoupled, then a particular choice of $\mathbf{O}$ is implied. Thus, from (1.36), with $\epsilon_{ji}^{\text{or}} = 0$, unless $i = j$, and on dropping the superscript $^{\text{or}}$ for simplicity, the $m$ uncoupled equations become

$$\hat{h}\psi_i = \epsilon_{ii}\psi_i \qquad (i = 1, 2, \ldots m) \qquad (1.38)$$

Each molecular orbital therefore satisfies a Schrödinger-like equation with the same effective Hamiltonian operator $\hat{h}$.

The dependence of $\hat{h}$ upon the unknown $\psi_i$ is an unavoidable feature of the model, and the pseudo-eigenvalue equations (1.38) must be solved iteratively by first guessing a set of $\psi_i$, in order to construct $\hat{h}$, and then solving (1.38) for the improved $\psi_i$. The whole process is repeated until self-consistency is reached.

Unfortunately, this method of solution, which yields the Hartree–Fock orbitals $\psi_i$, is feasible only for atoms, where the $\psi_i$ are obtained in numerical form. In the case of molecules, and sometimes now even for atoms,[26] a further approximation is made whereby each one-electron function, $\psi_i$, is expanded in terms of a linear combination of known one-electron functions—usually atomic orbitals. This is known as the LCAO approximation:

$$\psi_i = \sum_{\alpha=1}^{N} c_{\alpha i}\phi_\alpha \qquad i = 1, 2, 3 \ldots m \qquad (1.39)$$

(for an atom, of course, all $\phi_\alpha$ are centred on the same nucleus).

In principle, the expansion (1.39) ought to be over a complete set of one-electron functions, $\phi_\alpha$; but in practice this situation is never achieved, and the $\psi_i$ are usually expanded in terms of a limited set of $\phi_\alpha$ centred on each nucleus in the molecule. The $\phi_\alpha$ are often taken as simple Slater-type orbitals, but other choices of function can always be made[27] (see also Vol. 1). The energy obtained from molecular orbitals defined in terms of the complete set of expansion functions is known as the Hartree–Fock limit. This limit can be approached, in practice, by progressively increasing the number of (linearly independent) expansion functions in (1.39). However, as the $\psi_i$ are now constructed from an incomplete set of expansion functions, it is essential to

† Sometimes termed the canonical molecular orbitals.

determine the optimum $c_{\alpha i}$ which lead to a minimization of the molecular energy.

As a first step, therefore, the energy expression (1.32) must be rewritten to display the dependence of $E$ upon the undetermined coefficients $c_{\alpha i}$. This is readily accomplished by substituting (1.39) into (1.32). The permitted first-order change in the energy, arising from the variations $c_{\alpha i} \to c_{\alpha i} + \delta c_{\alpha i}$ in the LCAO coefficients, is then given by

$$\delta E = 4 \sum_{i,\,\alpha} \delta c_{\alpha i} \int \phi_\alpha \hat{h} \left( \sum_\alpha c_{\gamma j} \phi_\gamma \right) \mathrm{d}\mathbf{r}_1 - 4 \sum_{j,\,i} \epsilon_{ji} \sum_\alpha \delta c_{\alpha i} \int \phi_\alpha \left( \Sigma\, c_{\gamma j} \phi_\gamma \right) \mathrm{d}\mathbf{r}_1$$

which is required to vanish; that is,

$$\int \phi_\alpha \left[ \hat{h} \psi_i - \sum_j \epsilon_{ji} \psi_j \right] \mathrm{d}\mathbf{r}_1 = 0 \qquad \begin{array}{l} i = 1, 2, \ldots m \\[4pt] \alpha = 1, 2, \ldots N. \end{array}$$

$\hat{h}$ has the same form as the operator introduced earlier, except that the $\psi_i$ are now given in terms of a restricted LCAO expansion (no confusion should arise over the use of the same symbols for Hartree–Fock and LCAO defined quantities).

Thus, on introducing the LCAO expansion for $\psi_i$ in the above equations, and denoting the matrix elements of $\hat{h}$ and unity by $h_{\gamma\alpha}$ and $S_{\gamma\alpha}$, respectively, the optimum $c_{\alpha i}$ are found to satisfy

$$\sum_\gamma h_{\alpha\gamma} c_{\gamma i} - \sum_{j,\,\gamma} S_{\alpha\gamma} c_{\gamma j} \epsilon_{ji} = 0 \qquad \begin{array}{l} i = 1, 2, \ldots m \\[4pt] \alpha = 1, 2, \ldots N. \end{array} \tag{1.40}$$

These equations can be collected into the more manageable matrix equation

$$\mathbf{hc} = \mathbf{Sc}\boldsymbol{\epsilon} \tag{1.41}$$

Now the matrix $\mathbf{c}$ is not unique because, as seen earlier, the total energy is invariant to an orthogonal transformation of the $\psi_i$. If the transformed LCAO one-electron orbitals are collected into the row matrix $\boldsymbol{\psi}^{\mathrm{or}}$, then

$$\boldsymbol{\psi}^{\mathrm{or}} = \boldsymbol{\psi}\mathbf{O} = \boldsymbol{\phi}\mathbf{c}\mathbf{O} = \boldsymbol{\phi}\mathbf{c}^{\mathrm{or}}$$

where the penultimate step follows from the use of the matrix form of (1.39). Consequently, if the energy is minimized with respect to arbitrary changes in the $c_{\alpha i}^{\mathrm{or}}$, the above analysis shows that the optimum LCAO coefficients are now determined by

$$\mathbf{hc}^{\mathrm{or}} = \mathbf{Sc}^{\mathrm{or}}\boldsymbol{\epsilon}^{\mathrm{or}},$$

where it should be noted that, as expected, the elements of $\mathbf{h}$ are invariant to the transformation, as they depend upon the coefficients $c_{\alpha i}$ only in the form of elements of the matrix $(\mathbf{c}\ \mathbf{c}')$. This is best seen by expressing a typical matrix element of $\mathbf{h}$ in terms of integrals involving the basic expansion functions

$$
h_{\gamma\alpha} = \int \phi_\gamma(1)\hat{h}(1)\phi_\alpha(1)\,\mathrm{d}\mathbf{r}_1
$$

$$
= \int \phi_\gamma(1)\left[\hat{f}_1 + \sum_s (2\hat{J}_s(1) - K_s(1))\right]\phi_\alpha(1)\,\mathrm{d}\mathbf{r}_1
$$

$$
= f_{\gamma\alpha} + \sum_s \left[2\int\int \frac{\phi_\gamma(1)\psi_s(2)\phi_\alpha(1)\psi_s(2)}{r_{12}}\,\mathrm{d}\mathbf{r}_1\,\mathrm{d}\mathbf{r}_2 \right.
$$

$$
\left. - \int\int \frac{\phi_\gamma(1)\psi_s(2)\psi_s(1)\phi_\alpha(2)}{r_{12}}\,\mathrm{d}\mathbf{r}_1\,\mathrm{d}\mathbf{r}_2\right]
$$

$$
= f_{\gamma\alpha} + \sum_{\eta,\xi} P_{\eta\xi}[(\gamma\eta/\alpha\xi) - \tfrac{1}{2}(\gamma\eta/\xi\alpha)]
$$

$$
= f_{\gamma\alpha} + G_{\gamma\alpha} \tag{1.42}
$$

where

$$
(\gamma\eta/\alpha\xi) = \int \phi_\gamma(1)\phi_\eta(2)(1/r_{12})\phi_\alpha(1)\phi_\xi(2)\,\mathrm{d}\mathbf{r}_1\,\mathrm{d}\mathbf{r}_2
$$

and

$$
P_{\eta\xi} = 2(\mathbf{c}\,\mathbf{c}')_{\eta\xi} = 2\sum_k c_{\eta k}(\mathbf{c}')_{k\xi} = 2\sum_k c_{\eta k}c_{\xi k} = P_{\xi\eta},
$$

in the case of real coefficients.

The above matrix equation therefore reduces to the form

$$
\mathbf{hc} = \mathbf{ScO}\,\epsilon^{\mathrm{or}}\mathbf{O}'
$$

which identifies $\epsilon^{\mathrm{or}}$ with $\mathbf{O}'\epsilon\mathbf{O}$. Thus, just as in the earlier discussion of the derivation of the equations determining the Hartree–Fock orbitals, the transformation $\mathbf{O}$ is chosen in such a manner that $\epsilon^{\mathrm{or}}$ is diagonal: the corresponding orbitals $\psi_i^{\mathrm{or}}$ are the required LCAO molecular orbitals. As before, the superscript $^{\mathrm{or}}$ can now be dropped and the equations determining the optimum LCAO coefficients can be collected into the matrix equation

$$
\mathbf{hc} = \mathbf{Sc}\epsilon
$$

where $\epsilon$ is a diagonal matrix.

A convenient expression for the molecular electronic energy now follows from the use of (1.42):

$$
E = 2\sum_i \int \psi_i\hat{h}_\alpha\psi_i\,\mathrm{d}\mathbf{r}_1
$$

$$
= 2\sum_{i,\alpha,\gamma} c_{\gamma i}c_{\alpha i}h_{\gamma\alpha} - \sum_{i,\alpha,\gamma} c_{\gamma i}c_{\alpha i}G_{\gamma\alpha}
$$

$$
= \sum_{\alpha,\gamma} P_{\gamma\alpha}h_{\gamma\alpha} - \tfrac{1}{2}\sum_{\alpha,\gamma} P_{\gamma\alpha}G_{\gamma\alpha}
$$

$$
= \mathrm{tr}\,(\mathbf{Ph}) - \tfrac{1}{2}\,\mathrm{tr}\,(\mathbf{PG})
$$

$$
= \mathrm{tr}\,(\mathbf{Pf}) + \tfrac{1}{2}\,\mathrm{tr}\,(\mathbf{PG}).
$$

Although this expression does not involve the overlap matrix explicitly, the determination of **P** requires the solution of equations (1.40) which do contain **S**. It is of some interest, therefore, to examine the possibility of working with orthogonal atomic orbitals—preferably those obtained from the symmetric orthogonalization procedure as given in (1.21).

If the molecular orbitals are expanded in terms of the orthogonal orbitals, $\chi_\lambda$:

$$\psi_i = \sum_\alpha c_{\alpha i} \sum_\lambda S_{\lambda\alpha}^{1/2} \chi_\lambda = \sum_\lambda c_{\lambda i}^o \chi_\lambda \qquad (i = 1, 2, \ldots m) \qquad (1.43)$$

where

$$c_{\lambda i}^o = \sum_\alpha S_{\lambda\alpha}^{1/2} c_{\alpha i} = (\mathbf{S}^{1/2}\mathbf{c})_{\lambda i},$$

the expansion of (1.32) yields

$$E^o = \operatorname{tr} (\mathbf{P}^o \mathbf{h}^o) - \tfrac{1}{2} \operatorname{tr} (\mathbf{P}^o \mathbf{G}^o)$$

for the molecular energy. The superscript $o$ is added as a reminder that the various matrices are defined with respect to the set of orthogonalized atomic orbitals.

The two sets of atomic orbitals and LCAO coefficients are related by the transformations $\chi = \phi \mathbf{S}^{-1/2}$ and $\mathbf{c} = \mathbf{S}^{-1/2}\mathbf{c}^o$, respectively; and it is not difficult to verify that the two sets of **h**, **G**, and **P** matrices are related in the following manner:

$$\mathbf{P} = 2\mathbf{c}\mathbf{c}' = 2\mathbf{S}^{-1/2}(\mathbf{c}^o)(\mathbf{c}^o)'\mathbf{S}^{-1/2} = \mathbf{S}^{-1/2}\mathbf{P}^o\mathbf{S}^{-1/2},$$

$$\mathbf{h} = \mathbf{f} + \mathbf{G} = \mathbf{S}^{1/2}\mathbf{f}^o\mathbf{S}^{1/2} + \mathbf{S}^{1/2}\mathbf{G}^o\mathbf{S}^{1/2} = \mathbf{S}^{1/2}\mathbf{h}^o\mathbf{S}^{1/2}.$$

The energy expression therefore becomes

$$E^o = \operatorname{tr} (\mathbf{S}^{1/2}\mathbf{P}\mathbf{S}^{1/2}\mathbf{S}^{-1/2}\mathbf{h}\mathbf{S}^{-1/2}) - \tfrac{1}{2} \operatorname{tr} (\mathbf{S}^{-1/2}\mathbf{G}\mathbf{S}^{-1/2}\mathbf{S}^{1/2}\mathbf{P}\mathbf{S}^{1/2})$$

$$= \operatorname{tr} (\mathbf{Ph}) - \tfrac{1}{2} \operatorname{tr} (\mathbf{GP})$$

$$= E,$$

and an important difference emerges between the simple valence bond and molecular orbital theories: in molecular orbital theory, the calculation of the electronic energy is unaffected by orthogonalizing the basic atomic orbitals, in direct contradistinction to simple valence bond theory. This feature makes the molecular orbital method much more amenable for performing practical calculations. For on introducing the set of symmetrically orthogonalized atomic orbitals, (1.41) is transformed into

$$(\mathbf{S}^{1/2}\mathbf{h}^o\mathbf{S}^{1/2})(\mathbf{S}^{-1/2}\mathbf{c}^o) = \mathbf{S}(\mathbf{S}^{-1/2}\mathbf{c}^o)\boldsymbol{\epsilon}$$

which, after premultiplying by $\mathbf{S}^{-1/2}$, yields the matrix equation

$$\mathbf{h}^o\mathbf{c}^o = \mathbf{c}^o\boldsymbol{\epsilon} \quad \text{or} \quad (\mathbf{h}^o - \boldsymbol{\epsilon})\mathbf{c}^o = \mathbf{O} \qquad (1.44)$$

for determining the LCAO coefficients. It should be noticed that the $\epsilon_{ii}$, usually termed orbital energies, are unaffected by the transformation, since

$$\epsilon_{ii} = \int \psi_i \hat{h} \psi_i \, d\mathbf{r}_1 = \sum_{\alpha, \gamma} c_{\alpha i} c_{\gamma i} h_{\alpha\gamma} = (\mathbf{c}'\mathbf{h}\mathbf{c})_{ii}$$

$$= (\mathbf{c}^{o\prime}\mathbf{S}^{-1/2}\mathbf{h}\mathbf{S}^{-1/2}\mathbf{c}^o) = (\mathbf{c}^{o\prime}\mathbf{h}^o\mathbf{c}^o)_{ii}$$

$$= \epsilon_{ii}^o.$$

If $\mathbf{c}_i$ and $\mathbf{c}_i^o$ denote column vectors containing the LCAO coefficients of the $i$th doubly occupied molecular orbital, with respect to ordinary and orthogonalized atomic orbitals, respectively, then $\mathbf{c}$ (or $\mathbf{c}_i^o$) may be written in the form (see Hall[10])

$$\mathbf{c} = (\mathbf{c}_1\mathbf{c}_2\mathbf{c}_3 \ldots \mathbf{c}_m),$$

thereby enabling $\epsilon_{ii}$ to be written as

$$(\mathbf{c}'\mathbf{h}\mathbf{c})_{ii} = \left[ \begin{pmatrix} \mathbf{c}_1' \\ \mathbf{c}_2' \\ \cdot \\ \cdot \\ \cdot \\ \mathbf{c}_m' \end{pmatrix} \mathbf{h}(\mathbf{c}_1\mathbf{c}_2 \ldots \mathbf{c}_m) \right]_{ii}$$

$$= \mathbf{c}_i'\mathbf{h}\mathbf{c}_i = \mathbf{c}_i^{o\prime}\mathbf{h}^o\mathbf{c}_i^o.$$

The basic equations (1.41) and (1.44) therefore admit of the alternative descriptions
$$\mathbf{h}\mathbf{c}_i = \epsilon_{ii}\mathbf{S}\mathbf{c}_i$$
or $$\qquad\qquad (i = 1, 2, \ldots m) \qquad\qquad (1.45)$$
$$\mathbf{h}^o\mathbf{c}_i^o = \epsilon_{ii}\mathbf{c}_i^o,$$

respectively, depending upon the choice of the LCAO expansion functions. The latter equations are easier to solve because of the absence of the overlap matrix, $\mathbf{S}$. The solution proceeds by first constructing $\mathbf{h}^o$ from an initial set of guessed $\mathbf{c}_i^o$. The resulting linear homogeneous equations (1.45) then have a non-trivial solution if $|\mathbf{h}^o - \epsilon_{ii}\mathbf{I}| = 0$ : a determinantal equation which yields $N$ values for $\epsilon_{ii}$ if there are $N$ atomic functions in the expansion (1.39). The $m$ lowest energy values of $\epsilon_{ii}$, or more simply $\epsilon_i$, correspond to the energies of the molecular orbitals occupied in the ground state. The remaining $N-m$ orbital energies are associated with the (virtual) molecular orbitals which are unoccupied in the ground state; and their form is determined by the orthogonality requirements on the $N$ molecular orbitals. For each value of $\epsilon_{ii}$, the corresponding $\mathbf{c}_i^o$ is obtained by solving the secular equations in the usual way; the $\mathbf{c}_i$ are then given by $\mathbf{c}_i = \mathbf{S}^{-1/2}\mathbf{c}_i^o$.

The unoccupied orbitals are often used for describing excited electronic states. For example, the excitation of an electron from $\psi_p$ to $\psi_r$, which is unoccupied in the ground state, leads to singlet and triplet excited states, with wave functions

$$^{1,3}\Psi = N.\frac{1}{\sqrt{2}}\{\hat{A}\psi_1\bar{\psi}_1 \ldots \psi_p\bar{\psi}_r \ldots \psi_m\bar{\psi}_m \pm \hat{A}\psi_1\bar{\psi}_1 \ldots \bar{\psi}_p\psi_r \ldots \psi_m\bar{\psi}_m\}$$

and energies $\epsilon_r - \epsilon_p + 2K_{rp} - J_{rp}$ and $\epsilon_r - \epsilon_p - J_{rp}$, respectively, providing $\psi_p$ and $\psi_r$ are both non-degenerate. The positive sign is taken for the triplet wave function. Further states can also be constructed in which two or more electrons are excited into virtual orbitals.

An interesting situation now arises because some of the approximate excited state wave functions, produced by promoting electrons into virtual molecular orbitals, may interact with the (single configuration) approximate ground-state function if they possess the same spatial symmetry and spin multiplicity. However, some simplification results when the approximate wave functions are constructed out of solutions of (1.36); for, in this instance, there is no direct interaction of singly excited configurations with the ground state. This is readily seen by noticing that the matrix element

$$\int \Psi_0 \hat{H} \Psi_1 \, d\tau = \sqrt{2} \int \psi_p \hat{h} \psi_r \, d\mathbf{r}_1 = \sqrt{2}\,\epsilon_{pr} = \sqrt{2}\,\epsilon_{pp}\delta_{pr}$$

which connects the two states of interest is zero, because the $\psi_i$ are determined to ensure that $\epsilon$ is diagonal.

The interaction between the various approximate singlet states arises because the states themselves are not eigenfunctions of $\hat{H}$ (the approximate wave functions are, in fact, eigenfunctions of the effective Hamiltonian $\sum_i \hat{h}(i)$).†

An improved ground state wave function, $\Psi_0'$, is therefore obtained by allowing for the coupling between $\Psi_0$ and the higher energy approximate wave functions of the same symmetry (this is called configuration interaction):

$$\Psi_0' = \sum_{j=0} a_{j0}\Psi_j \tag{1.46}$$

In this instance, the $\Psi_j$ are singlet states arising from different configurations of electrons within the set of $N$ one-electron molecular orbitals. In the limit that all configurations are considered, (1.46) becomes identical with the corresponding expansion for $\Psi_0'$ in terms of all valence bond structures of the appropriate symmetry. The molecular orbital and valence bond theories are therefore equivalent in their respective limiting forms, and the only difference between them lies in the choice of the $2m$-electron expansion functions (see also Vol. 1). In practice, however, the two theories yield different results as the respective expansions for $\Psi_0'$ must be truncated in order to make the calculations tractable.

The error arising from the lack of correlation, implied by the use of the molecular orbital approximation to $\Psi_0$, can be reduced by allowing for the effects of configuration interaction (see (1.46)). But this procedure is often unsatisfactory from a computational point of view, as the description of

† See footnote p. 8, where $\sum_i \hat{h}(i)$ can be identified with $\hat{H}_1 + \hat{V}$.

excited states in terms of excitations to virtual orbitals leads to a slow convergence of (1.46). This follows because the virtual orbitals, which are determined only by the orthogonality requirements, are unlikely to have their maximum amplitudes in regions of space which are important for improving the degree of electron correlation.

A compromise solution to this problem is to limit the number of configurations in the expansion (1.46) to $\Psi_0$ and doubly excited configurations $\Psi_j(p{\rightarrow}r; p{\rightarrow}r)$ only. The variation theorem can then be used to optimize the energy with respect to the LCAO and configuration mixing coefficients. This procedure will yield the best average forms for the occupied molecular orbitals as well as the configuration mixing coefficients.[28-29] The method is particularly useful when poor results are obtained for the molecular binding energy from the single configuration wave function $\Psi_0$.   For example, $F_2$ is predicted to have a negative binding energy ($-$ 0·219 aJ) for the single configuration ground state; but the inclusion of only one suitably chosen doubly excited configuration[30] leads to a binding energy of 0·086 aJ, in comparison with the experimental value of 0·269 aJ. The same kind of improvement is also found[31] in similar calculations on $Li_2$: the binding energies calculated from $\Psi_0$ and from $\Psi_0'$, including six doubly excited configurations, are 0·027 aJ and 0·158 aJ, respectively, compared with the experimental value of 0·165 aJ.

Although the multi-configuration approach is desirable from an energetic viewpoint, the distribution of electronic charge is not expected to be altered appreciably by the inclusion of the doubly excited configurations (see Section 1.8). This is fortunate, as the determination of $\Psi_0$ alone is often an involved problem for a large molecule.

The calculation of $\Psi_0$ also yields, as a by-product, the set of $N$ orbital energies, which can be used for discussing the number, and nature, of possible electronic transitions. Obviously the use of virtual orbitals represents a limiting factor in accurate calculations but, more often than not, the complexity of the problem precludes anything but a rough order of magnitude calculation. The $\epsilon_{ii}$ can also be used for estimating molecular ionization potentials.

Consider the energy of the positive ion, formed by removing an electron from the doubly occupied molecular orbital, $\psi_t$:

$$E_+ = 2 \sum_{j(\neq t)} f_{jj} + f_{tt} + \sum_{i,j(\neq t)} (2J_{ij} - K_{ij}) + \sum_{k(\neq t)} (2J_{kt} - K_{kt})$$

The ionization potential is given by

$$\begin{aligned} E_+ - E &= 2 \sum_{j(\neq t)} f_{jj} + f_{tt} + \sum_{i,j(\neq t)} (2J_{ij} - K_{ij}) + \sum_{k(\neq t)} (2J_{kt} - K_{kt}) + J_{tt} \\ &\quad - 2 \sum_{j(\neq t)} f_{jj} - 2f_{tt} - \sum_{i,j(\neq t)} (2J_{ij} - K_{ij}) - 2 \sum_{k(\neq t)} (2J_{kt} - K_{kt}) \\ &= -f_{tt} - \sum_{k} (2J_{kt} - K_{kt}) \\ &= - \epsilon_{tt}. \end{aligned}$$

The derivation of this result (Koopmans' theorem)† requires there to be no relaxation of the charge distribution on ionization: that is, the same molecular orbitals are used for describing the electronic structure of both the neutral and the ionized molecules. The use of this theorem seems to yield remarkably good results, but this is thought to arise from a fortuitous cancellation of errors.[32] Apart from the real effects of relaxation, which must occur in the ion—presumably the electron distribution is contracted relative to the neutral molecule—there will be a concomitant change in the correlation energy (the correlation energy is usually defined as the difference between the Hartree–Fock energy, and the exact non-relativistic energy). The obvious method of improving the calculation is to perform the appropriate self-consistent open-shell calculation for the ion. But this approach sometimes leads to worse results than those obtained by use of Koopmans' theorem, because simple molecular orbital theory, by its very nature, cannot describe adequately the change in correlation energy on ionization (in the application of Koopmans' theorem, this change in correlation energy usually cancels the energy of reorganization). Only very accurate calculations on the atom or molecule and the corresponding ion, which include correlation effects explicitly in the actual wave function, can yield reliable values for the ionization potentials; for example, the calculations of Boys and Handy[33] on neon predict a lowest energy ionization potential of 3·457 aJ, compared with the "experimental" value of 3·448 aJ. This should be contrasted with a difference of 3·178 aJ in the self-consistent field total energies for Ne and Ne⁺, and with the value of 3·705 aJ as obtained by use of Koopmans' theorem.

## 1.6.   A brief résumé of open-shell molecular orbital theory

Up to this point the discussion has been limited to molecules possessing a ground state with a closed-shell electron configuration. For systems where the ground state arises from an open shell electron configuration, the determination of the ground-state wave function, $\Psi_0'$, which usually consists of a linear combination of determinants (say $t$ in number), presents a number of difficulties which are still incompletely resolved.

In the "unrestricted" (Hartree–Fock) theory,[34–35] the variation theorem is applied to the energy expression for only one of the $t$ antisymmetrized products of occupied spin orbitals (a single determinant), say

$$\Psi_0 = N\hat{A}\psi_1\bar{\varphi}_1\psi_2\bar{\varphi}_2 \ldots \psi_f \ldots \psi_p \tag{1.47}$$

where electrons with different components of spin angular momentum are allocated to different spatial orbitals; and $\psi_f \ldots \psi_p$ are occupied by electrons with $m_s = +\frac{1}{2}$ (α spin).

It is hoped that the optimum orbitals, obtained by minimizing the energy of

† Notice that Koopmans' theorem holds even for non-canonical molecular orbitals.

(1.47), can be used for constructing the fully symmetry adapted linear combination of determinants, even though $\Psi_0$ itself is neither an eigenfunction of $\hat{S}^2$, nor of any of the symmetry operators contained in $G$. Unfortunately, though, the $\alpha$- and $\beta$-spin orbitals are determined by different effective Hamiltonian operators, neither of which is, in general, invariant under the full set of symmetry operations contained in $G$. Thus, unlike the closed-shell molecule, where there is no loss of generality in constraining the molecular orbitals to transform like the irreducible representations of $G$ (the matrix $\mathbf{U}$ which diagonalizes the effective Hamiltonian, $\hat{h}$, of the closed shell problem, also (block) diagonalizes the matrices forming the reducible representation of $G$. Hence, for any $\hat{R}$ in $G$ (see footnote p. 69)

$$\hat{R}(\hat{h}\psi_i) = \hat{h}(\hat{R}\psi_i) = \epsilon_i(\hat{R}\psi_i)$$

where

$$\hat{R}\psi_i = \sum_{j=1}^{n} D_\Gamma(\hat{R})_{ji}\psi_j$$

and $n$ is the dimension of the irreducible representation $\Gamma$, and $\psi_j$ are the partner functions to $\psi_i$), it represents an additional constraint if the optimum one-electron orbitals are taken to be symmetry adapted. However, even if this symmetry constraint is imposed, $\Psi_0$ still represents a mixture of spin multiplets (i.e. it is not an eigenfunction of $\hat{S}^2$). This problem is usually resolved by annihilating unwanted spin multiplets by repeated application of the operator $\hat{O}_S = [\hat{S}^2 - S(S+1)]$, for selected values of $S$.† For example, in the simple case of the helium atom, the open-shell wave function

$$\Psi_0' = N\hat{A}\psi_s(1)\bar{\varphi}_s(2)$$

corresponds to a mixture of singlet and triplet wave functions (only two spin multiplets can arise from the coupling of two electron spins). The singlet wave function is therefore found by annihilating the triplet component with $\hat{O}_1$:

$$\hat{O}_1\Psi_0' = N\hat{A}\psi_s(1)\varphi_s(2)\hat{O}_1\alpha(1)\beta(2)$$
$$= N\hat{A}\psi_s\varphi_s[\hat{S}_+\hat{S}_- + \hat{S}_z^2 - \hat{S}_z]\alpha\beta$$
$$= -N\hat{A}\psi_s\bar{\varphi}_s + N\hat{A}\bar{\psi}_s\varphi_s.$$

A much more desirable approach is to apply the variation theorem to the fully symmetry adapted linear combination of determinantal wave functions derived from $\Psi_0$. In this "extended" procedure,[36–38] the one-electron orbitals are not required to be orthogonal or symmetry adapted; but, in fact, it is common practice to reduce the complexity of the calculations by constraining the one-electron orbitals to transform like the irreducible representations of $G$. Even with this symmetry constraint, the calculations are still very complex, and it is not surprising to find that applications of the method are limited to systems containing a small number of electrons.

† Unless $\Psi_0$ is associated with spatial degeneracy, in which case further annihilations are required to produce a properly symmetry adapted function.

Finally, there are the various "restricted" methods,[39-40] of which the Roothaan[39] method is probably the most popular—but not necessarily the best (see Nesbet[35]). In the Roothaan method, the variation theorem is applied to the expression for the average energy of all component wave functions possessing the same energy. The optimum one-electron orbitals are then determined by one effective Hamiltonian operator, which also displays the full symmetry of the nuclear framework (this is ensured by the energy averaging procedure). Thus, in this approach, the one-electron orbitals can be classified according to the irreducible representations of the molecular symmetry group. The main deficiency with the method (apart from the difficulty often experienced in achieving self-consistency) is the loss of the result that matrix elements of $\hat{H}$ between the ground state and excited states, arising from the excitation of a single electron, are required to vanish.[41] Also Koopmans' theorem no longer obtains.

The first deficiency is particularly important because, in addition to the extra effort expended in performing the open shell calculation, it is now necessary to allow for the coupling of $\Psi_0$ with singly excited states in order to ensure comparable accuracy of the results with those of the closed-shell calculations.

A further discussion of open-shell theory is now deferred until some applications are discussed in Chapter 3. Meanwhile, the discussion of molecular orbital theory is continued with an examination of the problems involved with developing a semi-empirical theory for closed-shell molecules.

## 1.7. Semi-empirical molecular orbital theory

The relative simplicity of the molecular orbital theory, when based on the use of a single configuration wave function, has led to its widespread use in recent years for obtaining information about the electronic structure of molecules. However, applications of the theory to large polyatomic molecules are severely hindered by the practical problem of computing vast numbers of two-electron integrals: for $N$ atomic orbitals the number of such integrals is of the order of $N^4$. This problem has caused a considerable amount of effort to be directed towards the production of a satisfactory semi-empirical theory. The main aim of a semi-empirical theory is to simulate the matrix elements of $\hat{h}$ by ad hoc approximations, or assumptions, which usually involve incorporating atomic spectral data into the calculation in some form.

An early attempt at a semi-empirical calculation on an inorganic molecule (see Vol. 2 for a full discussion on conjugated hydrocarbons) was made by Wolfsberg and Helmholtz[42] on the permanganate ion, $MnO_4^-$: a system for which a complete calculation presents monumental difficulties. Wolfsberg and Helmholtz approximated the diagonal elements of $\hat{h}$ by suitably chosen atomic ionization potentials; and off-diagonal elements were evaluated according to the prescription

$$h_{ij} = KS_{ij}(h_{ii} + h_{jj})$$

where $K$ is a parameter to be determined. This approximation essentially involves the use of (1.52), and any errors are hopefully absorbed in the constant $K$. Unfortunately, calculations made with this form of the theory are inadequate for a number of reasons, some of which are as follows:

(i) The intra-atomic part of $h_{ii}$ is given more correctly by the charge dependent orbital electronegativity (the arithmetic mean of an appropriate ionization potential and electron affinity), than by an ionization potential alone.

(ii) No cognizance is taken of the Coulomb potentials in $\hat{h}$ arising from the non-uniformity of the charge distribution around neighbouring atoms —a correction which is thought to be very large.[43]

(iii) The results are not invariant to the choice of coordinate system.[45]

(iv) Although different values of $K$ have been proposed, only $K = \frac{1}{2}$ is strictly valid, as otherwise the results are dependent on the choice for the zero of energy. The use of other values of $K$ requires the introduction of yet another parameter to ensure invariance with respect to changes in the zero of energy.[44]

(v) The total energy cannot be calculated, and hence nothing can be deduced about the molecular geometry.

(vi) Any assignment of observed spectral bands on the basis of electron jumps between one-electron energy levels is precluded, because of the neglect of all electron repulsion and exchange integrals: for example, the singly excited singlet and triplet states are always degenerate. In addition, no allowance can be made for configuration interaction, as this effect also depends upon the neglected two-electron integrals.

Criticism (iii) has been emphasized only fairly recently,[45] and invalidates the conclusions of any semi-empirical calculation made before 1965. This invariance requirement is particularly important, and a brief discussion of the basic principles is now given.

It was shown previously that the molecular energy is invariant to an orthogonal transformation on the basic set of atomic orbitals. There are two kinds of orthogonal transformation which are pertinent to the present discussion: first, those corresponding to rotations of local coordinate axes. These transformations only mix orbitals of the same $l$, and can be regarded as inducing the inverse transformation on the basic set of functions, with the axis system fixed (see McWeeny[46]). Secondly, there are orthogonal transformations like (1.16) which mix orbitals of different $l$. Both kinds of transformation can be represented by the introduction of the new set of atomic orbitals, $\chi_j$,

$$\chi_j = \sum_k O_{kj}\phi_k.$$

A typical element of the $\mathbf{f}$ matrix, with respect to the new set of atomic orbitals, is therefore given by

$$f_{pq}^0 = \int \chi_p(1)\hat{f}_1\chi_q(1)\,\mathrm{d}\mathbf{r}_1 = \sum_k \sum_l O_{kp}O_{lq}\int \phi_k(1)\hat{f}_1\phi_l(1)\,\mathrm{d}\mathbf{r}_1$$

$$= \sum_{k,l} O_{kp}f_{kl}O_{lq} = (\mathbf{O}'\mathbf{fO})_{pq};$$

$\mathbf{G}$ displays the same transformation properties as $\mathbf{f}$, since

$$\sum_j (2\hat{J}_j - \hat{K}_j)$$

is invariant to transformations of the kind under consideration, and so $\mathbf{h} \to \mathbf{h}^0 = \mathbf{O}'\mathbf{hO}$.

The significance of this result is that if approximations are used in evaluating the elements of $\mathbf{h}$, then the above transformation property must be maintained. The Wolfsberg–Helmholtz recipe for evaluating the off-diagonal elements of $\mathbf{h}$ does not meet with this requirement.

If the main aim of a semi-empirical theory is to simulate the results of a more complete calculation, then the above discussion shows that these expectations are unlikely to be met: for in the parametization of the problem no cognizance is usually taken of the invariancy requirements.

A much more practical, and reliable, procedure is to eliminate the difficulties of integral evaluation by using integral approximations which maintain the invariancy requirements. Since three- and four-centre two-electron integrals cause the most trouble in their evaluation, it is highly desirable to find a suitable integral approximation which reduces these integrals to a sum of two-centre integrals. The most obvious candidate for a multi-centre integral approximation is the one proposed by Mulliken,[47] where each overlap density is replaced by a simple sum of atom densities:

$$\phi_\alpha(1)\phi_\gamma(1) = \frac{S_{\alpha\gamma}}{2}[\phi_\alpha(1)\phi_\alpha(1) + \phi_\gamma(1)\phi_\gamma(1)].$$

A typical three-centre integral then becomes a linear combination of two-centre integrals

$$\int \phi_\alpha(1)\phi_\gamma(2)(1/r_{12})\phi_\mu(1)\phi_\gamma(2)\,\mathrm{d}\mathbf{r}_1\,\mathrm{d}\mathbf{r}_2 = \frac{S_{\alpha\mu}}{2}[(\alpha\gamma/\alpha\gamma) + (\mu\gamma/\mu\gamma)],$$

where $\phi_\alpha$, $\phi_\mu$ and $\phi_\gamma$ are centred on different atomic nuclei. Although this integral approximation is particularly simple to apply, it suffers from the limitation that it does not satisfy the invariancy requirements. However, integral approximations suggested by Löwdin[48] and by Newton and his co-workers[49] do have the necessary properties of rotational invariance, and calculations made on a limited number of molecules show that the results are in close agreement with those obtained from more accurate calculations.[50–51]

c

This gives great hope to the possibility that fairly reliable molecular orbital wave functions will be obtained soon for molecules which, at present, defy a completely non-empirical approach.

All of the discussion so far has been concerned more with the calculation of molecular electronic energies, than with any other molecular properties. However, the function describing the distribution of electronic charge density is of considerable chemical interest: for a knowledge of this function is valuable for discussing the nature of the bonding in a particular molecule, or for comparing changes in the distribution of the bonding electrons in similar molecules. The calculation of this function is also important, because it should be possible to compare the results, in appropriate instances, with those deduced from X-ray diffraction studies on molecular crystals (after allowing for the effects of thermal motion)—providing, of course, that the crystal density can be partitioned reliably into components associated with individual molecules. But even when this situation does not obtain, a comparison with the X-ray results may give useful information about the changes in the electron distribution which occur on crystal formation.

## 1.8.   The electron density function

The wave function for a molecule containing $2m$ electrons is a function of $6m$ spatial coordinates. However, the function describing the distribution of electronic charge density relates to the probability per unit volume of finding an electron at a given point in space: that is, it depends upon only three spatial coordinates.

The probability of finding electron 1 in volume element $d\tau_1$ (i.e. with spatial variables between $r_1$ and $r_1 + dr_1$, and spin variable between $s_1$ and $s_1 + ds_1$), electron 2 in volume element $d\tau_2$, etc., is given by

$$\Psi(1, 2, \ldots 2m)\Psi^*(1, 2, \ldots 2m) \, d\tau_1 \, d\tau_2 \ldots d\tau_{2m}. \qquad (1.48)$$

The probability of finding electron 1 in $d\tau_1$, and electrons 2 to $2m$ with any values of the space and spin coordinates is obtained by summing the various probabilities (1.48), keeping the coordinates of electron 1 fixed:

$$\left( \int \Psi(1, 2, \ldots 2m)\Psi^*(1, 2, \ldots 2m) \, d\tau_2 \ldots d\tau_{2m} \right) d\tau_1.$$

The indistinguishability of electrons ensures that any other electron will have the same probability of being found in the chosen volume element. Hence, the probability of finding an electron in $d\tau_1$ at the point $r_1$ with spin variable $s_1$ is just

$$2m \left( \int \Psi(1, 2, \ldots 2m)\Psi^*(1, 2, \ldots 2m) \, d\tau_2 \ldots d\tau_{2m} \right) d\tau_1 \qquad (1.49)$$

which is usually written as $\rho(1, 1)\, d\tau_1$, The probability of finding an electron in the spatial volume element $d\mathbf{r}_1$, irrespective of spin component, is therefore

$$\left(\int \rho(1, 1)\, ds_1\right) d\mathbf{r}_1 \equiv P(1, 1)\, d\mathbf{r}_1$$

where $P(1, 1)\, d\mathbf{r}_1$ consists of the sum of the probabilities that the electron in $d\mathbf{r}_1$ has spin $\alpha$ or spin $\beta$.

The form of $P(1, 1)$ is now derived for two choices of molecular wave function: first, the orthogonal pair wave function, and then the molecular orbital wave function with limited configuration interaction.

The orthogonal pair wave function is given by the antisymmetrized product of pair functions

$$\Psi_0 = N2^{m/2}\hat{A}\psi_1(1, 2)\psi_2(3, 4) \ldots \psi_m(2m - 1, 2m) \times$$
$$\times \alpha(1)\beta(2) \ldots \alpha(2m - 1)\beta(2m)$$

and, on substitution into (1.49), yields

$$\rho(1, 1) = (2m).N^2.2^m.\int \left(\sum_{\hat{P}} (-1)^P \hat{P}\psi_1 \ldots \psi_m\alpha(1) \ldots \beta(2m)\right) \times$$
$$\times \left(\sum_{\hat{Q}} (-1)^Q \hat{Q}\psi_1 \ldots \psi_m\alpha(1) \ldots \beta(2m)\right) d\tau_2 \ldots d\tau_{2m}.$$

The integrations in this expression are best performed by following arguments similar to those used in obtaining (1.27):

$$\rho(1, 1) = (2m)N^2.2^m(2m - 1)! \times$$
$$\times \left\{\sum_j \int \psi_j(1, 2)\psi_j(1, 2)[\alpha(1)\alpha(1)\beta(2)\beta(2) + \beta(1)\beta(1)\alpha(2)\alpha(2)]\, d\tau_2\right\}$$
$$= N^2.2^m.(2m)! \int \left(\sum_j \psi_j(1, 2)\psi_j(1, 2)\right) \times$$
$$\times (\alpha(1)\alpha(1)\beta(2)\beta(2) + \beta(1)\beta(1)\alpha(2)\alpha(2))\, d\tau_2.$$

The square of the normalization constant is readily evaluated as $1/\{2^m.(2m)!\}$ by using the electron conservation requirement $\int \rho(1, 1)\, d\tau_1 = 2m$. Thus, after eliminating $N^2$ from the expression for $\rho(1, 1)$, and integrating over the spin coordinates of electron 1,

$$P(1, 1) = 2\sum_j \int \psi_j(1, 2)\psi_j(1, 2)\, d\mathbf{r}_2.$$

On introducing an expansion like (1.20) for the $\psi_j$, and then integrating over the coordinates of electron 2, the expression for $P(1, 1)$ reduces to a simple sum of weighted products of the orthogonal atomic orbitals. For example, the pair function

$$\psi_j(1, 2) = c_{1j}[\chi_\alpha(1)\chi_\gamma(2) + \chi_\gamma(1)\chi_\alpha(2)] + c_{2j}\chi_\alpha(1)\chi_\alpha(2) + c_{3j}\chi_\gamma(1)\chi_\gamma(2)$$

gives rise to the following contribution to $P(1, 1)$:

$$
\begin{aligned}
2 \int \psi_j(1, 2)\psi_j(1, 2)\, d\tau_2 &= 2c_{1j}^2[\chi_\alpha(1)\chi_\alpha(1) + \chi_\gamma(1)\chi_\gamma(1)] + 2c_{2j}^2\chi_\alpha(1)\chi_\alpha(1) \\
&\quad + 2c_{3j}^2\chi_\gamma(1)\chi_\gamma(1) + 4c_{1j}c_{2j}\chi_\gamma(1)\chi_\alpha(1) \\
&\qquad\qquad\qquad\qquad\qquad + 4c_{1j}c_{3j}\chi_\alpha(1)\chi_\gamma(1) \\
&= P_{\alpha\alpha}^0\chi_\alpha(1)\chi_\alpha(1) + P_{\gamma\gamma}^0\chi_\gamma(1)\chi_\gamma(1) \\
&\quad + P_{\alpha\gamma}^0\chi_\alpha(1)\chi_\gamma(1) + P_{\gamma\alpha}^0\chi_\gamma(1)\chi_\alpha(1).
\end{aligned}
$$

The summation of these contributions, over all pair functions, then leads to a $P(1, 1)$ of the form

$$
P(1, 1) = \sum_{\mu,\,\lambda} P_{\mu\lambda}^0 \chi_\mu(1)\chi_\lambda(1),
$$

where, it should be remembered, in this version of the pair theory, a given pair of atomic functions only appears in one pair function. $P(1, 1)$ can also be expressed in terms of the non-orthogonal atomic functions (these functions still contain small contributions of core functions though) by making use of the transformation (1.21):

$$
\begin{aligned}
P(1, 1) &= \sum_{\eta,\,\xi,\,\mu,\,\lambda} S_{\eta\mu}^{-1/2} P_{\mu\lambda}^0 S_{\lambda\xi}^{-1/2} \phi_\eta(1)\phi_\xi(1) \\
&= \sum_{\eta,\,\xi} P_{\eta\xi}\phi_\eta(1)\phi_\xi(1).
\end{aligned}
$$

The molecular orbital wave function, allowing for configuration interaction, leads to the general expression

$$
P(1, 1) = \sum_{k,\,l} c_{k0}c_{l0} \int \Psi_k(1, 2, \ldots 2m)\Psi_l(1, 2, \ldots 2m) \times
$$

$$
\times\, d\tau_2 \ldots d\tau_{2m}\, ds_1 \quad (1.50)
$$

which reduces to

$$
c_{00}^2 \int \Psi_0\Psi_0\, d\tau_2 \ldots d\tau_{2m}\, ds_1 + \sum_k{}' c_{k0}^2 \int \Psi_k\Psi_k\, d\tau_2 \ldots d\tau_{2m}\, ds_1
$$

if the terms in the summation are restricted to $\Psi_0$ and doubly excited configurations only. The terms with $k \neq l$ disappear because of the orthogonality of the basic molecular orbitals. This is readily verified by considering the contribution of the term arising from the product of the approximate ground state function, $\Psi_0$, and any doubly excited state, $\Psi_l$:

$$
\begin{aligned}
&\int \Psi_0\Psi_l\, d\tau_2 \ldots d\tau_{2m}\, ds_1 \\
&= \frac{1}{(2m)!} \int \left( \sum_{\hat P} (-1)^P \hat{P}\psi_1(1)\bar\psi_1(2) \ldots \psi_p(x)\bar\psi_p(y) \ldots \right) \times \\
&\qquad\qquad \times \left( \sum_{\hat Q} (-1)^Q \hat{Q}\psi_1(1)\bar\psi_1(2) \ldots \psi_r(x)\bar\psi_r(y) \ldots \right) d\tau_2 \ldots d\tau_{2m}\, ds_1
\end{aligned}
$$

Maximum matching of the orbital products can only be achieved if $\hat{P}$ and $\hat{Q}$ correspond to the same permutation. But even with this choice of permutations, there are still two spin-orbital mismatches, and integration over the coordinates of $2m - 1$ electrons can only yield zero. Hence, only terms with $k = l$ survive in (1.50), and give a contribution to $P(1, 1)$ of the form

$$\int \Psi'_l \Psi_l \, d\tau_2 \ldots d\tau_{2m} \, ds_1 = 2 \sum_i \psi_i(1)\psi_i(1),$$

where $i$ runs over all doubly occupied orbitals in $\Psi_l$. An equation of this form arises from each non-vanishing contribution to $P(1, 1)$, and so the overall result can be written as

$$P(1, 1) = \sum_i n_i \psi_i(1)\psi_i(1), \tag{1.51}$$

where $n_i$ is the effective occupation number of orbital $\psi_i$. Clearly, for a single configuration molecular orbital calculation, each of the occupied molecular orbitals has an occupation number of two: but the inclusion of configuration interaction causes slight deviations in these occupation numbers, as orbitals not occupied in $\Psi_0$ now contribute to (1.51). It should also be noted that the expression (1.51) for $P(1, 1)$ is only diagonal, with respect to the $\psi_i$ functions, if singly, triply, . . . excited configurations are excluded from the expansion for $\Psi'_0$.

In the simplest situation, which is the one most commonly experienced, $\Psi_0$ is approximated by the lowest energy configuration of $m$ doubly occupied molecular orbitals. The expansion for $P(1, 1)$ then takes the form

$$P(1, 1) = 2 \sum_{i=1}^{m} \psi_i(1)\psi_i(1) = 2 \sum_i \sum_{\alpha, \gamma} c_{\alpha i} c_{\gamma i} \phi_\alpha(1)\phi_\gamma(1)$$
$$= \sum_{\alpha, \gamma} P_{\alpha\gamma} \phi_\alpha(1)\phi_\gamma(1)$$

which only differs from the corresponding expressions for the orthogonal pair, or valence bond, wave functions, in the values assigned to the coefficients $P_{\alpha\gamma}$.

The one-electron density function, $P(1, 1)$, can therefore be evaluated at various points in space, providing the coefficients $P_{\alpha\gamma}$ are known (the functional forms of the basic atomic orbitals are fixed at the start of the appropriate calculation). The practical difficulties in displaying the variation of $P$ in three dimensions are usually overcome by considering its variation in selected plane sections—in the case of a linear molecule, for example, it is convenient to select a plane containing the nuclei. As examples of these plots, the contours of $P(1, 1)$, or $P(\mathbf{r}_1)$, are shown in Fig. 1.4 for CO and $N_2$. The polarization of the charge distribution in CO is clearly visible. In this instance, the three-dimensional form for each $P$ function is obtained by rotating the appropriate plot about the internuclear axis.

Although diagrams of the kind shown in Fig. 1.4 give a good indication of the overall distribution of electronic charge, it would be very useful if this distribution could be subdivided into component parts associated with different regions of the molecule: the variation in the electron density of a particular atom, bond or lone-pair region could then be studied in different molecules. There are, in fact, two general methods currently in use for discussing the subdivision of the molecular electronic charge distribution.

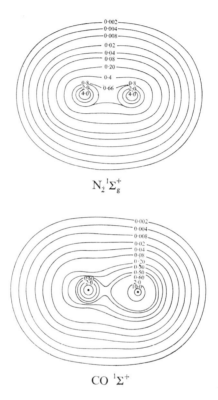

$N_2\,{}^1\Sigma_g^+$

CO $^1\Sigma^+$

FIG. 1.4. Contours of the one-electron density function, $P(\mathbf{r}_1)$, in a plane containing the nuclei, for the ground electronic states of the nitrogen and carbon monoxide molecules (these figures are reproduced by from *J. Chem. Phys*, 1968, **49**, 1653 by courtesy of the authors, R. F. W. Bader and A. D. Bandrauk). The units of $P$ are electrons/a.u.³

First, following the pioneering work of Mulliken,[52] the charge distribution can be discussed in terms of the numbers $P_{\alpha\gamma}$ and $S_{\alpha\gamma}$. Alternatively, suitable difference density functions can be calculated by subtracting the appropriate atomic $P(1, 1)$ functions, situated at their respective site in the molecule, from the molecular $P(1, 1)$ function. The overall changes in electronic charge density, as a result of molecule formation, should then become apparent.

In the Mulliken approach, the coefficients, $P_{\alpha\alpha}$, in the expansion for $P(1, 1)$, are taken as a measure of the amount of charge associated with the atomic orbital $\phi_\alpha$. Similarly, $2P_{\alpha\gamma}S_{\alpha\gamma}$ gives a measure of the amount of overlap charge between the two nuclei associated with the atomic orbitals $\phi_\alpha$, $\phi_\gamma$. For a pair of chosen atoms $p$, $q$, Mulliken defined net atom and bond populations by

$$n(p) = \sum_{\alpha \text{ on } p} P_{\alpha\alpha}, \qquad n(p, q) = \sum_{\substack{\alpha, \text{ on } p \\ \gamma \text{ on } q}} 2P_{\alpha\gamma}S_{\alpha\gamma}$$

where positive and negative values of $n(p, q)$ are taken to indicate a bonding or antibonding situation between the two atoms; that is, there is a net attraction or repulsion, respectively, leading to charge accumulation or depletion. It should be noted that, unlike the individual orbital and overlap densities, the net populations are invariant to local orthogonal transformations—such as a rotation of the coordinate system on one of the atoms. The $n(p)$ and $n(p, q)$ are not invariant to transformations which mix atomic orbitals on different atoms: such as occurs, for example, in any orthogonalization procedure.

Since the conservation of electronic charge density requires

$$\sum_\alpha P_{\alpha\alpha} + 2 \sum_{\alpha < \gamma} P_{\alpha\gamma}S_{\alpha\gamma} = \text{tr } (\mathbf{PS}) = 2m$$

the sum of the $n(p)$, over all atoms in the molecule, does not equal $2m$. The numbers, $n(p)$, are therefore not suited to discussing charge distributions in terms of formal charges which can be associated with the various atoms in the molecule. Mulliken attempted to circumvent this difficulty by introducing the concept of a gross atom population, $N(p)$. These quantities are obtained by first equi-partitioning the overlap populations between the pair of atoms concerned, and then summing the appropriate contributions on each atom:

$$N(p) = n(p) + \tfrac{1}{2} {\sum_q}' n(p, q)$$

It is apparent that the equi-partitioning of the overlap populations implies the use of the approximation

$$\phi_\alpha(1)\phi_\gamma(1) \sim \tfrac{1}{2}S_{\alpha\gamma}[\phi_\alpha(1)\phi_\alpha(1) + \phi_\gamma(1)\phi_\gamma(1)] \tag{1.52}$$

to eliminate the overlap densities in the expression for $P(1, 1)$. That is,

$$P(1, 1) = \sum_\alpha \left( P_{\alpha\alpha} + {\sum_\gamma}' S_{\alpha\gamma}P_{\alpha\gamma} \right) \phi_\alpha(1)\phi_\alpha(1)$$

giving,

$$N(p) = \sum_{\alpha \text{ on } p} \left( P_{\alpha\alpha} + {\sum_\gamma}' S_{\alpha\gamma}P_{\alpha\gamma} \right) = n(p) + \tfrac{1}{2} {\sum_q}' n(p, q).$$

Gross atom charges, or effective atomic charges, are then defined by subtracting the gross atom population from the number of electrons provided by the neutral atom, $N_0(p)$:

$$Q(p) = N_0(p) - N(p)$$

where $Q(p)$ is given in units of $e$, the electronic charge.

The arbitrary division of the bond populations implied by the use of (1.52) is, to some extent, unsatisfactory. Although there are many ways of partitioning the overlap densities, the approach based on the use of (1.52) suffers from the disadvantage that the dipole moment of the overlap density, along the line joining the two nuclei, is not conserved: that is, the value of the dipole moment depends upon which side of (1.52) is used in its evaluation. For this reason, Löwdin[53] suggested an improved partitioning of the form

$$\phi_\alpha(1)\phi_\gamma(1) = S_{\alpha\gamma}[\lambda\phi_\alpha(1)\phi_\alpha(1) + (1 - \lambda)\phi_\gamma(1)\phi_\gamma(1)]$$

where $\lambda$ is chosen to ensure that the dipole moment is conserved. The calculated values of $Q(p)$ are now different from those obtained with the Mulliken approximation, (1.52), and the two sets of results are compared in Table 1.2.

TABLE 1.2

*The calculated effective atomic charges, $Q(p)$ (in units of e), as obtained from the Mulliken and the Löwdin partitionings of the atomic overlap densities*

| Molecule $XYZ$ | | HCN | FCN | LiH | NH | FH | CO |
|---|---|---|---|---|---|---|---|
| $R_{XY}$(nm) | | 0·106 | 0·126 | 0·159 | 0·104 | 0·092 | 0·113 |
| $R_{YZ}$(nm) | | 0·116 | 0·116 | — | — | — | — |
| Mulliken partitioning | $Q(X)$ | +0·36 | −0·12 | +0·31 | −0·08 | −0·15 | +0·11 |
| | $Q(Y)$ | −0·14 | +0·26 | −0·31 | +0·08 | +0·15 | −0·11 |
| | $Q(Z)$ | −0·22 | −0·14 | — | — | — | — |
| Löwdin partitioning | $Q(X)$ | +0·14 | 0·00 | +0·62 | +0·22 | +0·10 | +0·14 |
| | $Q(Y)$ | +0·02 | −0·01 | −0·62 | −0·22 | −0·10 | −0·14 |
| | $Q(Z)$ | −0·16 | +0·01 | — | — | — | — |
| References | | 51[a] | 51[a] | 54, 55 | 54, 55 | 54, 55 | 54, 55 |

[a] Unpublished calculations by G. Doggett and G. Howat, using the PLA method as described in ref. 51.

It is extremely unlikely that the form of the three-dimensional electron density function, $P(1, 1)$, is adequately described by the set of numbers, $Q(p)$. This follows, because it is often difficult to justify the allocation of the various orbital charge densities to particular atoms: for example, if $\phi_\alpha$ is a diffuse atomic orbital, with maximum amplitude in the region of another atom, $q$ say, then it seems only natural that the atom $q$ should have some share of the orbital charge, here represented by $P_{\alpha\alpha}$. But this is never allowed in the existing forms of population analysis. The magnitudes, and signs, of one particular set of atomic charges are therefore of little significance (see Table

1.2), and it is imperative to use effective atomic charges for comparing trends in the charge distributions of similar molecules. A further discussion of the problems inherent in performing a population analysis will be found in a recent paper by Mulliken.[56]

Most population analyses have been made for mloecules whose approximate ground-state wave functions have been found by means of molecular orbital theory. However, as seen above, a similar population analysis can be made for an orthogonal pair or valence bond wave function—the only difference being that, for the same set of atomic orbitals, a different set of $P_{\alpha\gamma}$ coefficients results. Although the population analysis may have its limitations, the coefficients $P_{\alpha\gamma}$ can be used as a means of comparing the three kinds of molecular wave functions. Unfortunately, though, for historical reasons, a completely different way of handling valence bond wave functions was developed, because the theory was nearly always applied within an empirical framework to unsaturated hydrocarbons (see Vol. 2). In this particular form of the theory, the atomic orbitals were assumed to be orthogonal, and all ionic structures were neglected. The deficiencies arising from these assumptions are hopefully absorbed in the parameterization of the energy expression. But this does not alter the rather disturbing fact that all $P_{\alpha\gamma}$ coefficients are zero if $\alpha \neq \gamma$. It is not surprising, therefore, to find that notions pertaining to the variation in bond length with bond order had to be developed in terms of empirically defined bond orders, which have no physical significance. This should be contrasted with the situation in molecular orbital theory, where the corresponding coefficients do not vanish—even if overlap integrals are neglected. The non-vanishing of the $P_{\alpha\gamma}$ coefficients in the overlap neglected molecular orbital theory results from the implicit inclusion of ionic structures within the molecular orbital wave function. For example, in the case of the hydrogen molecule, the simple Heitler–London wave function, with overlap terms put equal to zero, gives rise to the following one-electron density function:

$$P(1, 1) = \phi_\alpha(1)\phi_\alpha(1) + \phi_\gamma(1)\phi_\gamma(1)$$

while the molecular orbital wave function gives rise to a density function of the form

$$P(1, 1) = \phi_\alpha(1)\phi_\alpha(1) + \phi_\gamma(1)\phi_\gamma(1) + \phi_\alpha(1)\phi_\gamma(1) + \phi_\gamma(1)\phi_\alpha(1).$$

Hence, the notion of bond order appears quite naturally in a simple molecular orbital theory, but has no relation with the empirically defined valence bond bond orders. It is therefore most unfortunate that the concept a bond order–length relation has been developed in totally different ways in molecular orbital and valence bond theories. It would be much better to include the effects of orbital non-orthogonality in a legitimate way—but then the concept of bond order disappears, and is replaced by that of bond population.

The second method for examining molecular electronic charge distributions was originally suggested by Roux and her co-workers.[57] The technique is to subtract suitably chosen atomic densities, situated at the appropriate sites in the molecule, from the molecular density:

$$\Delta P(1, 1) = P_{\text{mol}}(1, 1) - \sum_{\text{atoms}} P_{\text{atom}}(1, 1).$$

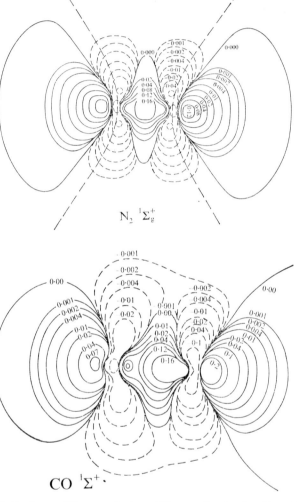

FIG. 1.5. Contours of the one-electron difference density function, $\Delta P(\mathbf{r}_1)$, in a plane containing the nuclei, for the ground electronic states of the nitrogen and carbon monoxide molecules (these figures are reproduced from *J. Chem. Phys.*, 1968, **49**, 1653 by courtesy of the authors, R. F. W. Bader and A. D. Bandrauk). The units of $\Delta P$ are electrons/a.u.$^3$

The form of the $\Delta P$ function usually shows, in a striking manner, how the electronic charge is redistributed on molecule formation. The method suffers only from the slight disadvantage that it is not always possible to select unambiguously the appropriate atomic electronic densities. Results for CO and $N_2$ (see Fig. 1.5) show that, in both cases, charge is removed from the vicinity of the nuclei, and redistributed in the regions normally ascribed to bond and lone pairs.

If the $\Delta P$ maps are to be of any use, then their form must be insensitive to the choice of atomic orbitals used in the calculations. This appears to be so, except in the special case of the hydrogen molecule. In this instance, the $\Delta P$ maps calculated from the valence bond and molecular orbital wave functions are virtually indistinguishable, but their form is sensitive to the choice of atomic $1s$ functions: the use of free atom and energy minimized orbital exponents ($k = 1{\cdot}0$ and $1{\cdot}166$, respectively), at the corresponding equilibrium nuclear separations of $1{\cdot}52a_0$ and $1{\cdot}41a_0$, respectively, results in very different $\Delta P$ plots as shown in Fig. 1.6 (only the results for the valence bond wave function are shown).

Now the virial theorem requires a specific balance between the average kinetic and potential energies for the molecular equilibrium nuclear configuration (see also Vol. 1), that is

$$2\bar{T}/\bar{V} = -1$$

where

$$\bar{T} = \int \Psi_0 \left[ -\tfrac{1}{2} \sum_i \nabla_i^2 \right] \Psi_0 \, d\tau \Big/ \int \Psi_0\Psi_0 \, d\tau,$$

$$\bar{V} = \int \Psi_0 \left[ -\sum_{i,\alpha} \frac{Z_\alpha}{r_{i\alpha}} + \sum_{i<j} \frac{1}{r_{ij}} + \sum_{\alpha<\beta} \frac{Z_\alpha Z_\beta}{R_{\alpha\beta}} \right] \Psi_0 \, d\tau \Big/ \int \Psi_0\Psi_0 \, d\tau.$$

Since this balance is obtained only in the calculation using contracted $1s$ atomic orbitals, the results in Fig. 1.6 (a) are clearly spurious (see also Table 1.3).

Fortunately, the effects of orbital contraction are usually much less marked in other molecules, as the availability of $1s$, $2s$ and $2p$ atomic orbitals (at least, as far as first-row atoms are concerned) allows for some flexibility in the molecular wave function. Part of the readjustment of the charge distribution, required to ensure the correct balance between the average kinetic and potential energies, is taken up by changes in the appropriate LCAO co-efficients in a molecular orbital calculation; or by changes in the relative weights of covalent and ionic structures in a valence-bond calculation. Hence, orbital exponent optimization is not expected to be so crucial in calculations on many-electron molecules—a conclusion which is amply supported by the values of $-2\bar{T}/\bar{V}$ obtained in calculations using only a limited set of free atom atomic orbitals—often Slater-type orbitals (see Table 1.3).†

† Except perhaps for molecules containing hydrogen atoms.

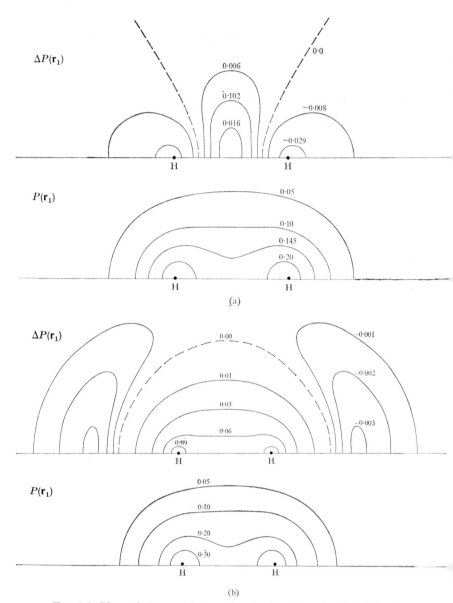

Fig. 1.6. Plots of the one-electron density functions $P$ and $\Delta P$ for the hydrogen molecule, as obtained from the simple Heitler–London wave function. The results in (a) are obtained with a choice of free atom $1s$ atomic orbitals ($k = 1\cdot0$), with the nuclei separated by $1\cdot52a_0$; the results in (b) are obtained from molecularly optimized $1s$ atomic functions ($k = 1\cdot166$), with an internuclear separation of $1\cdot41a_0$ (in both cases the chosen internuclear separation is the one which minimizes the molecular energy). The units of $P$ and $\Delta P$ are electrons/a.u.[3]

TABLE 1.3

*The calculated values of* $-2\bar{T}/\bar{V}$ *for some typical molecules*[a]

| Molecule | Molecular geometry (nm) | Calculation | Choice of atomic orbitals | $-2\bar{T}/\bar{V}$ | Refs. |
|---|---|---|---|---|---|
| $H_2$ | H $\overset{0\cdot080}{\rule{1cm}{0.4pt}}$ H | Heitler–London | Free-atom $1s$: exponent unity. | 0·860 | 58 |
| $H_2$ | H $\overset{0\cdot075}{\rule{1cm}{0.4pt}}$ H | Heitler–London | $1s$: exponent optimized ($k = 1\cdot166$) | 1·000 | 58 |
| HCN | H $\overset{0\cdot106}{\rule{1cm}{0.4pt}}$ C $\overset{0\cdot116}{\rule{1cm}{0.4pt}}$ N | Single configuration molecular orbital | Minimal basis set of one Slater-type function per atomic core or valence orbital: exponents optimized for the free atoms | 0·9934 | 59 |
| $N_3^-$ | N $\overset{0\cdot112}{\rule{1cm}{0.4pt}}$ N $\overset{0\cdot112}{\rule{1cm}{0.4pt}}$ N | As above | As above | 1·0025 | 59 |
| HNCO | H, 128° N $\overset{}{\underset{0\cdot121}{\rule{1cm}{0.4pt}}}$ C $\overset{0\cdot117}{\rule{1cm}{0.4pt}}$ O (0·098) | As above | As above | 0·9957 | 59 |
| $N_2O$ | N $\overset{0\cdot113}{\rule{1cm}{0.4pt}}$ N $\overset{0\cdot117}{\rule{1cm}{0.4pt}}$ O | As above | Gaussian lobe functions | 1·0003 | 60 |
| $F_2O$ | F, 103½° F, 0·141, 0·141, O | As above | Minimal basis of Slater-type orbitals: exponents chosen by Slater's rules | 1·0055 | 61 |
| $F_2O$ | As above | As above | Minimal basis of Slater-type orbitals: exponents optimized for the free atoms | 0·9978 | 61 |
| NaF | Na $\overset{0\cdot200}{\rule{1cm}{0.4pt}}$ F | As above | Minimal basis set of two Slater-type functions per atomic core or valence orbital | 0·9994 | 62 |
| NaF | As above | As above | As above, plus exponent optimized $3p$ and $3d$ Slater-type functions on Na and F | 1·00001 | 62 |
| NaF | As above | As above | Minimal basis set of Clementi (ref. 26) atomic functions plus exponent optimized $3p$, $3d$ and $4f$ Slater-type functions | 1·00012 | 62 |
| NaF | Na $\overset{0\cdot192}{\rule{1cm}{0.4pt}}$ F | As above | As above | 1·000015 | 62 |

[a] Strictly speaking, the ratio $-2\bar{T}/\bar{V}$ should only be calculated for the minimum energy geometry (this was only achieved in the calculations on $H_2$ and the last calculation on NaF): any departure from unity then indicates the "goodness" of the atomic orbital basis set. The assumption of the experimental geometry in all other cases, except those indicated above, is likely to incur only a small error.

Although the use of optimized orbital exponents is clearly desirable, it is not so crucial as in the case of the hydrogen molecule, where there is only one atomic orbital centred on each nucleus. This is an important conclusion, for it means that the $\Delta P$ maps will be relatively insensitive to the choice of atomic orbital exponents (but this is unlikely to be true for the calculated atomic charges). However, it is not worth while, for example, to construct $\Delta P$ maps for molecular calculations based on the use of simple Slater-type orbitals, if Hartree–Fock orbitals are used for describing the free atoms; for the $\Delta P$ maps would then contain information about the poor quality of the LCAO expansion, in addition to details of the redistributed molecular electronic charge density. Clearly, it is advantageous to use the best free-atom atomic orbitals for the molecular calculation.

This concludes the preliminary, but rather extensive, discussion of the methods used in calculating the electronic structure of molecules, and the remainder of the book is concerned with the application of these methods to inorganic molecules.

## REFERENCES

1. PEKERIS, C. L., *Phys. Rev.*, 1959, **115**, 1216.
2. KOŁOS, W. and WOLNIEWICZ, L., *J. Chem. Phys.*, 1968, **49**, 404.
3. BORN, M. and OPPENHEIMER, J. R., *Ann. Phys.*, 1927, **84**, 457.
4. JAHN, H. A. and TELLER, E., *Proc. Roy. Soc.*, 1937, A **161**, 220.
5. LONGUET-HIGGINS, H. C., ÖPIK, H. C., PRYCE, M. H. L. and SACK, R. A., *Proc. Roy. Soc.*, 1958, A **244**, 1.
6. LONGUET-HIGGINS, H. C., *Adv. Spect.*, 1961, **2**, 429.
7. JAMES, H. M. and COOLIDGE, A. S., *J. Chem. Phys.*, 1933, **1**, 825.
8. HARRIS, F. E., *J. Chem. Phys.*, 1960, **32**, 3.
9. BISHOP, D. M., *Advances in Quantum Chemistry*, 1967, **3**, 25.
10. HALL, G. G., *Matrices and Tensors* (Topic 1, volume 4 of this series), Pergamon Press, Oxford (1963).
11. KING, H. F., STANTON, R. E., KIM, H., WYATT, R. E. and PARR, R. G., *J. Chem. Phys.*, 1967, **47**, 1936.
12. PROSSER, F. and HAGSTROM, S., *Intern. J. Quantum Chem.*, 1968, **2**, 89.
13. SERBER, R., *J. Chem. Phys.*, 1934, **2**, 697.
14. SEITZ, F. and SHERMAN, A., *J. Chem. Phys.*, 1934, **2**, 11.
15. LÖWDIN, P.-O., *J. Chem. Phys.*, 1950, **18**, 365.
16. McWEENY, R., *Proc. Roy. Soc.*, 1954, A **223**, 63, 306.
17. SLATER, J. C., *J. Chem. Phys.*, 1951, **19**, 220.
18. KLESSINGER, M. and McWEENY, R., *J. Chem. Phys.*, 1965, **42**, 3343.
19. STUART, J. D. and HURST, R. P., *Mol. Phys.*, 1965, **9**, 265.
20. KLESSINGER, M., *Chem. Phys. Lett.*, 1969, **4**, 144.
21. KLESSINGER, M., *Faraday Soc. Symp.*, 1968, **2**, 73.
22. FRANCHINI, P. F. and VERGANI, C., *Theor. Chim. Acta*, 1969, **13**, 46.
23. COULSON, C. A., *Valence* (2nd ed.), Oxford University Press (1961).
24. DIRAC, P. A. M., *Proc. Roy. Soc.*, 1929, A **123**, 714.
25. HALL, G. G. and LENNARD-JONES, J. E., *Proc. Roy. Soc.*, 1950, A **202**, 155.
26. CLEMENTI, E., *Tables of Atomic Functions*, Supplement to *I.B.M. J. Res. Dev.*, 1965, **9**, 2.
27. HALL, G. G., *Rep. Prog. Phys.*, 1959, **22**, 1.
28. HINZE, J. and ROOTHAAN, C. C. J., Supplement 40, *Prog. Theor. Phys.*, 1967, **37**.

29.  VEILLARD, A. and CLEMENTI, E., *Theor. Chim. Acta*, 1967, **7**, 133.
30.  DAS, G. and WAHL, A. C., *J. Chem. Phys.*, 1966, **44**, 87.
31.  DAS, G., *J. Chem. Phys.*, 1967, **46**, 1568.
32.  CLEMENTI, E., *J. Chem. Phys.*, 1967, **47**, 4485.
33.  BOYS, S. F. and HANDY, N. C., *Proc. Roy. Soc.*, 1969, A **310**, 63.
34.  LÖWDIN, P.-O., *Rev. Mod. Phys.*, 1963, **35**, 496.
35.  NESBET, R. K., *Rev. Mod. Phys.*, 1961, **33**, 28.
36.  LÖWDIN, P.-O., *Phys. Rev.*, 1955, **97**, 1509.
37.  JUCYS, A. P., KAMMSKAS, V. A. and KAVECKIS, V. J., *Intern. J. Quantum Chem.*, 1968, **2**, 405.
38.  HAMEED, S., HUI, S. S., MUSHER, J. I. and SCHULMAN, J. M., *J. Chem. Phys.*, 1969, **51**, 502.
39.  ROOTHAAN, C. C. J., *Rev. Mod. Phys.*, 1960, **32**, 179.
40.  NESBET, R. K., *Proc. Roy. Soc.*, 1955, A **230**, 312.
41.  CARLSON, K. D. and WHITMAN, D. R., *Intern. J. Quantum Chem.*, 1967, **1S**, 81.
42.  WOLFSBERG, M. and HELMHOLTZ, L., *J. Chem. Phys.*, 1952, **20**, 837.
43.  JØRGENSEN, C. K., HORNER, S. M., HATFIELD, W. E. and TYREE, S. Y., *Intern. J. Quantum Chem.*, 1967, **1**, 191.
44.  BERTHIER, G., DEL RE, G. and VEILLARD, A., *Nuovo Cimento*, 1966, **44**, 315.
45.  POPLE, J. A., SANTRY, D. P. and SEGAL, G. A., *J. Chem. Phys.*, 1965, **43**, S129.
46.  McWEENY, R., *Symmetry—An Introduction to Group Theory and its Applications* (Topic 1, volume 3 of this series), Pergamon Press, Oxford (1963).
47.  MULLIKEN, R. S., *J. Chim. Phys.*, 1949, **46**, 497.
48.  LÖWDIN, P.-O., *Phil. Mag. Supplement*, 1956, **5**, 1.
49.  NEWTON, M. D., OSTLAND, N. S. and POPLE, J. A., *J. Chem. Phys.*, 1968, **51**, 912.
50.  NEWTON, M. D., *J. Chem. Phys.*, 1969, **51**, 3917.
51.  DOGGETT, G. and McKENDRICK, A., *Faraday Soc. Symp.*, 1968, **2**, 32.
52.  MULLIKEN, R. S., *J. Chem. Phys.*, 1955, **23**, 1833, 1841, 2338, 2343.
53.  LÖWDIN, P.-O., *J. Chem. Phys.*, 1953, **31**, 374.
54.  RANSIL, B. J., *Rev. Mod. Phys.*, 1960, **32**, 245.
55.  DOGGETT, G., *J. Chem. Soc.*, 1969, A, 229.
56.  MULLIKEN, R. S., *J. Chem. Phys.*, 1962, **36**, 3428.
57.  ROUX, M., BESAINOU, S. and DAUDEL, R., *J. Chim. Phys.*, 1956, **54**, 218.
58.  HIRSCHFELDER, J. O. and KINCAID, J. F., *Phys. Rev.*, 1937, **52**, 658.
59.  BONACCORSI, R., PETRONGOLO, C., SCROCCO, E. and TOMASI, J., *J. Chem. Phys.*, 1968, **48**, 1500.
60.  PEYERIMHOFF, S. D. and BUENKER, R. J., *J. Chem. Phys.*, 1968, **49**, 2473.
61.  PETRONGOLO, C., SCROCCO, E. and TOMASI, J., *J. Chem. Phys.*, 1968, **48**, 407.
62.  MATCHA, R. L., *J. Chem. Phys.*, 1967, **47**, 5295.

# BIBLIOGRAPHY

HERZBERG, G., The dissociation energy of the hydrogen molecule, *J. Mol. Spectroscopy*, 1970, **33**, 147.

MOFFITT, W. E. and LIEHR, A. D., Configurational instability of degenerate electronic states, *Phys. Rev.*, 1957, **106**, 1195.

PARKS, J. M. and PARR, R. G., Theory of separated electron pairs, *J. Chem. Phys.*, 1958, **28**, 335.

HURLEY, A. C., LENNARD-JONES, J. E. and POPLE, J. A., The molecular orbital theory of chemical valency: XVI. A theory of paired electrons in polyatomic molecules, *Proc. Roy. Soc.*, 1953, A **220**, 446.

KOTANI, M., OHNO, K. and KAYAMA, K., Quantum mechanics of electronic structure of simple molecules, *Handbuch der Physik*, 1961, 37/2, Molecules II, Springer-Verlag, Berlin.

SLATER, J. C., *Quantum Theory of Molecules and Solids*, 1963 (volume I), McGraw-Hill Book Co. Inc., New York.

ROOTHAAN, C. C. J., New developments in molecular orbital theory, *Rev. Mod. Phys.*, 1951, **23**, 29.

BLINDER, S. F., Basic concepts of self-consistent-field theory, *Amer. J. Phys.*, 1965, **33**, 431.

NESBET, R. K., Electron correlation in atoms and molecules, *Adv. Chem. Phys.*, 1965, **IX**, 321.

WAHL, A. C., BERTONCINI, P. J., DAS, G. and GILBERT, T. L., Recent progress beyond the Hartree–Fock method for diatomic molecules: the method of optimized valence configurations, *Intern. J. Quantum Chem.*, 1967, **1S**, 123.

HARRIS, F. E., Molecular Orbital Theory, *Adv. Quantum Chem.*, 1967, **3**, 61.

BADER, R. F. W., HENNEKER, W. H. and CADE, P. E., Molecular charge distributions and chemical binding, *J. Chem. Phys.*, 1967, **46**, 3341.

RANSIL, B. J. and SINAI, J. J., Toward a charge density analysis of the chemical bond; the charge density bond model, *J. Chem. Phys.*, 1967, **46**, 4050.

WAHL, A. C., Molecular orbital densities—pictorial studies, *Science*, 1966, **151**, 961.

# CHAPTER 2

# THE ELECTRONIC STRUCTURE OF SOME MOLECULES CONTAINING A CENTRAL SECOND-ROW ATOM

## 2.1. Introduction

A discussion of the bonding, and associated stereochemical features, in molecules containing first-row atoms has already been given in Vol. 1: electronic-structure calculations on these molecules are straightforward in principle, and there are many results available which can be used for developing a quantitative theory of the bonding. The situation is less well defined for molecules containing a central second-row atom because, unlike the corresponding first-row molecules, the second-row atom may be in a state of high valence: for example, unlike oxygen, sulphur can form stable hexa- and tetra-covalent compounds with fluorine. The causes of this expansion of the valence shell, as it is often called, which is typified by so many second-row atoms, presents a problem in electronic structure theory which is unlikely to be solved quantitatively in the foreseeable future. This should be contrasted with the situation in which the second-row atom is in a state of low valence, where the problems are not very different from those involved in describing the electronic structure of the corresponding first-row molecules. The present chapter is therefore restricted to a discussion of those molecules in which the central second-row atom is in a state of high valence. Particular emphasis is placed on the structure of $SF_6$ but, towards the end of the chapter, the ground-state geometries of other sulphur- and phosphorus-containing compounds are also discussed.

The historical development of the subject stems from an apparently successful application of a simple valence bond theory by Pauling,[1] and independently by Slater,[2] to the problem of rationalizing the stereochemistry of coordination compounds—particularly those involving a transition metal ion. The main assumption of this model is in the augmentation of the occupied valence shell atomic orbitals by orbitals which are unoccupied in the ground state (the stereochemistries of molecules containing central first-row atoms or second-row atoms in low valence states are excluded from the discussion). These additional orbitals may have the same, or higher, principal quantum number as the valence shell atomic orbitals occupied in the ground state. As the method is concerned solely with the electron configuration of the central atom, or ion, it cannot be expected to offer any rationale, for example, about the relative abilities of fluorine and hydrogen to stabilize the hexa-covalent state of sulphur.

The Pauling–Slater model for discussing octahedrally coordinated compounds requires the construction of six geometrically equivalent hybrid atomic orbitals ($sp^3d^2$ hybrids) associated with the central atom, for use in bonding with the ligands. But before dealing with the construction of these directed atomic functions, which is a useful exercise in itself, it is first necessary to examine the symmetry properties of the various ordinary atomic orbitals (of different $l$ quantum number) in an environment of octahedral symmetry.

The original $(2l+1)$-fold degeneracy associated with each atomic orbital (with the exception of atomic orbitals in one-electron atoms, where the degeneracy is

$$\sum_{l=0}^{n-1} (2l + 1);$$

that is $n^2$) is unlikely to persist in an environment of reduced symmetry, since it is the assumption of the central field approximation (see Vol. 1) which leads to the existence of the usual degeneracies in atoms. For example, in an environment of regular octahedral symmetry the permitted maximum degeneracy is three (see the $O_h$ group character table in Appendix II). It therefore follows that the orbital degeneracies associated with values of $l \geqslant 2$, as well as the degeneracies associated with atomic many-electron states characterized by $L \geqslant 2$, will be partially removed when the atom in question is situated in a molecular environment of octahedral symmetry.

Molecular environments of octahedral and tetrahedral symmetries are associated with the presence of forty-eight and twenty-four symmetry

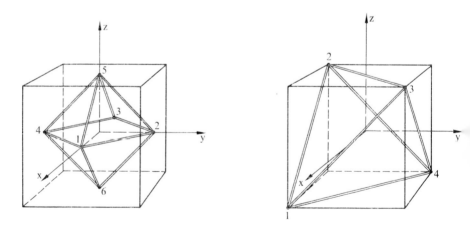

FIG. 2.1. Axis convention for compounds possessing octahedral and tetrahedral point group symmetries.

elements, respectively (see Fig. 2.1 for axis convention). The two sets of corresponding symmetry operations form the groups $O_h$ and $T_d$, respectively. In addition, the operations of $T_d$ form a sub-group of $O_h$:

$$O_h = (\hat{E} + \hat{J})T_d \equiv T_d + \hat{J}T_d:$$

that is, $O_h$ is formed from $T_d$ and its left-hand coset with respect to $\hat{J}$. $T_d$ itself can be generated from $T$, the group of pure rotations which transform the tetrahedron into itself (the group of rotations, $O$, which transform the cube and octahedron into themselves, is formed from $T$ and its left-hand coset with respect to $\hat{C}_4^z : O = T + \hat{C}_4^z T$):

$$T_d = (\hat{E} + \hat{J}\hat{C}_4^z)T.$$

The generators of $T$ (see McWeeny[3]) are $\hat{C}_2^z$ and $\hat{C}_3^{xyz}$, while the generators of $T_d$ and $O_h$ are $\hat{J}\hat{C}_4^z$, $\hat{C}_3^{xyz}$ and $\hat{C}_4^z$, $\hat{C}_3^{xyz}$, $\hat{J}$, respectively. All other group operations are obtained by taking suitable products of powers of the generators. It is therefore only necessary to investigate the effects of a small number of symmetry operations when attempting to classify functions according to the irreducible representations of the appropriate molecular point group. This procedure, which is now outlined for $s$, $p$, and $d$ atomic orbitals, requires knowledge of some standard group theoretical results, which can be found, for example, in the book by McWeeny.[3]

Let the value of the general atomic orbital function, $\phi$, at the point $\mathbf{r}$ be denoted by $\phi(\mathbf{r})$ where, as the notation implies, the origin of the coordinate system is taken on the atomic nucleus. The position vector $\mathbf{r}$ is further expressed in terms of the (orthogonal) Cartesian base vectors $\mathbf{e}_1$, $\mathbf{e}_2$, $\mathbf{e}_3$:

$$\mathbf{r} = x\mathbf{e}_1 + y\mathbf{e}_2 + z\mathbf{e}_3 = (\mathbf{e}_1\mathbf{e}_2\mathbf{e}_3)\begin{pmatrix} x \\ y \\ z \end{pmatrix}$$

The function $\phi$ is now transformed by applying the operator $\hat{R}$, of the appropriate point group, $G$. McWeeny shows that the value of the transformed function $\hat{R}\phi$ at the point $\mathbf{r}$ is equal to the value of $\phi$ at the backwards transformed point, $\hat{R}^{-1}\mathbf{r}$:†

$$\hat{R}\phi(\mathbf{r}) = \phi(\hat{R}^{-1}\mathbf{r}).$$

Thus, for $s$ atomic orbitals, where $\phi_s(\mathbf{r}) = \sqrt{(1/4\pi)}g_s(r)$, and $g_s(r)$ is the normalized radial part of $\phi_s$,

$$\hat{R}\phi_s(\mathbf{r}) = \phi_s(\hat{R}^{-1}\mathbf{r}) = \sqrt{(1/4\pi)}g_s(\hat{R}^{-1}r) = \sqrt{(1/4\pi)}g_s(r) = \phi_s(\mathbf{r})$$

† The symmetry operation $\hat{R}$ transforms points in coordinate space (or, more generally, in configuration space) according to $\mathbf{r} \to \mathbf{r}' = \hat{R}\mathbf{r}$. The operation describing the transformation of $\phi$ in function space is usually referred to as $\hat{P}_R$, so that $\hat{P}_R\phi(\mathbf{r}) = \phi(\hat{R}^{-1}\mathbf{r})$. However, the McWeeny[3] notation is used here, where it must be understood that $\hat{R}\phi$ is the name of the transformed function rather than $\hat{P}_R\phi$.

This result shows that $s$ atomic orbitals transform according to the totally symmetric irreducible representation of $G$; that is $a_1$ or $a_{1g}$ in the case of the $T_d$ or $O_h$ point groups, respectively (see Appendix II).

Now consider the set of $p$ atomic orbitals in an environment of tetrahedral symmetry. The $p_x$ orbital may be written in the form

$$\phi_x(\mathbf{r}) = \sqrt{(3/4\pi)}(x/r)\,g_p(r) = \sqrt{(3/4\pi)}(\mathbf{e}_1.\mathbf{r}).1/r.g_p(r)$$

which, under transformation by $\hat{R}$, yields

$$\hat{R}\phi_x(\mathbf{r}) = \phi_x(\hat{R}^{-1}\mathbf{r}) = \sqrt{(3/4\pi)}(\mathbf{e}_1.\hat{R}^{-1}\mathbf{r}).1/r.g_p(r)$$
$$= \sqrt{(3/4\pi)}(\hat{R}\mathbf{e}_1.\mathbf{r}).1/r.g_p(r)$$

The last step follows from the property

$$(\hat{R}\mathbf{a}.\hat{R}\mathbf{b}) = (\mathbf{a}.\mathbf{b}),$$

since $\hat{R}$ induces only orthogonal transformations.

The transformation properties of the $p$ atomic orbitals are therefore identical[3] to those of the triplet of base vectors, $\mathbf{e}_i$. But the transformed base vectors $\hat{R}\mathbf{e}_i$ can be expressed in terms of the original base vectors by

$$\hat{R}\mathbf{e}_i = \sum_{j=1}^{3} D(\hat{R})_{ji}\mathbf{e}_j,$$

where the collection of numbers $D(\hat{R})_{ji}$ forms the matrix representative of $\hat{R}$ with respect to the chosen set of basis functions (here $\mathbf{e}_i$; $i = 1, 2, 3$).

Let the generators $\hat{S}\hat{C}_4^z$, $\hat{C}_3^{xyz}$ for $T_d$ be denoted by $\hat{R}_1$, $\hat{R}_2$ (notice that $\hat{R}_2$ is also a generator for $O_h$). Application of $\hat{R}_2$ then induces the following transformation on the set of atomic $p$-orbitals:

$$\hat{R}_2\phi_x = \sqrt{(3/4\pi)}(\mathbf{e}_2.\mathbf{r}).1/r.g_p(r) = \phi_y,$$
$$\hat{R}_2\phi_y = \sqrt{(3/4\pi)}(\mathbf{e}_3.\mathbf{r}).1/r.g_p(r) = \phi_z,$$
$$\hat{R}_2\phi_z = \sqrt{(3/4\pi)}(\mathbf{e}_1.\mathbf{r}).1/r.g_p(r) = \phi_x.$$

That is,

$$\hat{R}_2\widehat{\phi_x\phi_y\phi_z} = \widehat{\phi_x\phi_y\phi_z} \begin{pmatrix} 0 & 0 & 1 \\ 1 & 0 & 0 \\ 0 & 1 & 0 \end{pmatrix}$$

which yields a character of 0. A similar procedure yields the character $-1$ for the generator $\hat{R}_1$. These results are sufficient to show that the set of three $p$ functions span the irreducible representation $t_2$ of $T_d$ (notice that lower-case letters are used to characterize the symmetries of one-electron functions). The characters for the remaining symmetry operations of $T_d$ are readily generated by forming the appropriate products of $\mathbf{D}(\hat{R}_1)$ and $\mathbf{D}(\hat{R}_2)$, and taking the diagonal sums of the resulting matrices. For example, since

$$\hat{E} = \hat{C}_3^{xyz}\hat{C}_3^{xyz}\hat{C}_3^{xyz},$$

the matrix representative $\mathbf{D}(\hat{E})$ is given by

$$\begin{pmatrix} 0 & 0 & 1 \\ 1 & 0 & 0 \\ 0 & 1 & 0 \end{pmatrix}\begin{pmatrix} 0 & 0 & 1 \\ 1 & 0 & 0 \\ 0 & 1 & 0 \end{pmatrix}\begin{pmatrix} 0 & 0 & 1 \\ 1 & 0 & 0 \\ 0 & 1 & 0 \end{pmatrix} = \begin{pmatrix} 1 & 0 & 0 \\ 0 & 1 & 0 \\ 0 & 0 & 1 \end{pmatrix}$$

with a character of three.

By using the same method for the generators of $O_h$, the three $p$ orbitals are found to provide a basis for the irreducible representation $t_{1u}$.

The situation for $d$ atomic orbitals is more interesting. The matrix representatives for $\hat{R}_1$ and $\hat{R}_2$ (of $T_d$) are given by

$$\mathbf{D}(\hat{R}_1) = \begin{pmatrix} 0 & 0 & 1 & 0 & 0 \\ 0 & -1 & 0 & 0 & 0 \\ 1 & 0 & 0 & 0 & 0 \\ 0 & 0 & 0 & -1 & 0 \\ 0 & 0 & 0 & 0 & 1 \end{pmatrix}, \text{ and } \mathbf{D}(\hat{R}_2) = \begin{pmatrix} 0 & 0 & 1 & 0 & 0 \\ 1 & 0 & 0 & 0 & 0 \\ 0 & 1 & 0 & 0 & 0 \\ 0 & 0 & 0 & -\frac{1}{2} & -\frac{1}{2}\sqrt{3} \\ 0 & 0 & 0 & \frac{1}{2}\sqrt{3} & -\frac{1}{2} \end{pmatrix}$$

with respect to the set of five $d$-orbitals $d_{xz}$, $d_{xy}$, $d_{yz}$, $d_{x^2-y^2}$, and $d_{z^2}$.

The set of twenty-four matrices, generated from $\mathbf{D}(\hat{R}_1)$ and $\mathbf{D}(\hat{R}_2)$, cannot form an irreducible representation of $T_d$, as the character for $\hat{E}$ is five. The representation must therefore be reducible, as no irreducible representation of this dimension is allowed (see Appendix II). In fact, it is very easy to see which irreducible representations are present because, for the particular choice of axes, the matrices $\mathbf{D}(\hat{R}_1)$ and $\mathbf{D}(\hat{R}_2)$ both have the same block-diagonal form. And since this block-diagonal form is maintained under matrix multiplication, all other matrix representatives formed from the generators will have the same block structure. The orbitals $d_{xz}$, $d_{xy}$, and $d_{yz}$ therefore span a representation of dimension three, while $d_{x^2-y^2}$ and $d_{z^2}$ span one of dimension two. The corresponding characters are $-1$, $0$ and $0$, $-1$ for $\hat{R}_1$ and $\hat{R}_2$, respectively; and examination of the character table for $T_d$ shows that the two sets of functions span the irreducible representations $t_2$ and $e$, respectively. The five-fold degeneracy of the free atom $d$-orbitals is therefore partially lifted in an environment of tetrahedral symmetry. But further calculations are still required in order to establish the magnitude of the energy separation between the two sets of states.

A similar analysis for $O_h$ shows that the $d$-orbital manifold splits into the same two groups of levels, but they are now labelled by the irreducible representations $t_{2g}$ and $e_g$.

The scheme whereby atomic orbitals are labelled by their associated values of $(l, m_l)$ is no longer adequate, as the potential experienced by the electrons deviates from spherical symmetry; labels corresponding to the irreducible representations of the appropriate molecular point group are used instead. For example, even though the degeneracy of the three atomic $p$-orbitals is not removed in an octahedral environment, they are still labelled by the irreducible representation $t_{1u}$ of $O_h$.

It is now possible to return to the problem of determining the form of the Pauling–Slater localized atomic functions which are used in describing the electronic structure of complexes possessing high symmetry.

The particular orbitals required for the formation of six geometrically equivalent hybrid atomic orbitals, each of which is directed preferentially along one of the positive or negative coordinate axes, are found most readily by using the symmetry properties of the octahedron (see Fig. 2.1 for the choice of axes). Each one of the forty-eight symmetry operations in $O_h$ gives rise to a permutation of the six hybrid atomic orbitals. For example, the transformation induced by the two-fold rotation $\hat{C}_4^{\prime z}$ is symbolized by

$$\hat{C}_4^{\prime z}\overbrace{o_1 o_2 o_3 o_4 o_5 o_6} = \overbrace{o_4 o_1 o_2 o_3 o_5 o_6}$$

where the six hybrids, $o_i$, are collected into a row matrix. The row matrix on the right-hand side of this equation can be written in terms of the original row matrix times a $6 \times 6$ matrix, which is the matrix representative of $\hat{C}_4^{\prime z}$ with respect to the chosen basis (hybrid orbitals, $o_i$):

$$\overbrace{o_4 o_1 o_2 o_3 o_5 o_6} = \overbrace{o_1 o_2 o_3 o_4 o_5 o_6} \begin{pmatrix} 0 & 1 & 0 & 0 & 0 & 0 \\ 0 & 0 & 1 & 0 & 0 & 0 \\ 0 & 0 & 0 & 1 & 0 & 0 \\ 1 & 0 & 0 & 0 & 0 & 0 \\ 0 & 0 & 0 & 0 & 1 & 0 \\ 0 & 0 & 0 & 0 & 0 & 1 \end{pmatrix}$$

The set of forty-eight matrices, one for each symmetry operation of $O_h$, forms a representation of the group. But as $O_h$ contains no irreducible representation of dimension greater than 3 (see Appendix II), the representation formed by the set of $6 \times 6$ matrices must be reducible. The decomposition of this reducible representation, using standard group theoretical methods, shows the presence of the $a_{1g}$, $t_{1u}$, and $e_g$ irreducible representations (see Vol. 1, or McWeeny,[3] for a discussion of the basic group theoretical techniques). It follows that atomic functions possessing these symmetries are required to construct six geometrically equivalent localized atomic functions: thus, on excluding functions with $l > 2$, the only functions which satisfy the symmetry requirements are $s(a_{1g})$, $p(t_{1u})$, and the pair of $d$-orbitals possessing maxima in their probability densities along the coordinate axes, $d_{z^2}$ and $d_{x^2-y^2}(e_g)$. The general localized function is therefore given by

$$o_i = a_1 s + a_2 p_x + a_3 p_y + a_4 p_z + a_5 d_{z^2} + a_6 d_{x^2-y^2}.$$

Now the hybrid $o_5$, directed along the positive $z$-axis, remains invariant under any of the permitted rotations around the $z$-axis ($C_4^z$, $\bar{C}_4^z$, $C_2^z$). This requires $a_2 = a_3 = a_6 = 0$. But as the relative weights of $s : p : d$ character in each hybrid are in the ratio $1 : 3 : 2$, the coefficients $a_4$ and $a_5$ are fixed as $\sqrt{3}.a_1$ and $\sqrt{2}.a_1$, respectively; the normalization requirement then determines $a_1$, and hence

$$o_5 = (1/\sqrt{6})(s + \sqrt{3}.p_z + \sqrt{2}.d_{z^2}).$$

The analytic expressions for the remaining $o_i$ are determined in the following way. If the local $z$-axis through the appropriate $o_i$ is denoted by $z'$, then

$$o_i = (1/\sqrt{6})(s+\sqrt{3}.p_{z'} + \sqrt{2}.d_{z'^2})$$

using exactly the same arguments as in the construction of $o_5$. The local axes must now be referred to the original set, in order that all atomic orbitals are defined with respect to the same coordinate system. Alternatively, $o_i$ may be regarded as the result of rotating $o_5$ into the position occupied by $o_i$. As an example, consider the transformation of $o_5$ under the generator $\hat{R}_2(\hat{C}_3^{xyz})$:

$$\hat{R}_2 o_5 = (1/\sqrt{6})(\hat{R}_2 s + \sqrt{3}.\hat{R}_2 p_z + \sqrt{2}.\hat{R}_2 d_{z^2})$$
$$= (1/\sqrt{6})(s + \sqrt{3}.\sqrt{(3/4\pi)}(\hat{R}_2 \mathbf{e}_3.\mathbf{r}).1/r.g_p(r)$$
$$+ \sqrt{2}.\sqrt{(5/16\pi)}[3(\hat{R}_2 \mathbf{e}_3.\mathbf{r})^2 - r^2].1/r^2.g_d(r))$$
$$= (1/\sqrt{6})(s + \sqrt{3}.\sqrt{(3/4\pi)}(\mathbf{e}_1.\mathbf{r}).1/r.g_p(r)$$
$$+ \sqrt{2}.\sqrt{(5/16\pi)}[3(\mathbf{e}_1.\mathbf{r})^2-r^2].1/r^2.g_d(r))$$
$$= (1\sqrt{6})(s + \sqrt{3}.p_x + \sqrt{2}.d_{x^2}).$$

The $d_{x^2}$ orbital must now be expressed in terms of the members of the linearly independent set of five $d$ atomic functions: in this instance, only $d_{z^2}$ and $d_{x^2-y^2}$ functions need be considered, as the remaining functions cannot contribute to a $d$-orbital used for $\sigma$-bonding. These functions have the forms

$$d_{z^2} = \sqrt{(5/16\pi)}(3\cos^2\theta - 1)g_d(r); \quad d_{x^2-y^2} = \sqrt{(15/16\pi)}\sin^2\theta\cos 2\phi.g_d(r)$$

and it is not difficult to show that

$$d_{x^2} = \sqrt{(5/16\pi)}.[3\sin^2\theta\cos^2\phi - 1]g_d(r) = -\tfrac{1}{2}d_{z^2} + (\sqrt{3}/2)\,d_{x^2-y^2}.$$

Hence,      $$\hat{R}_2 o_5 = (1/\sqrt{6})(s + \sqrt{3}.p_x + \sqrt{(3/2)}\,d_{x^2-y^2} - (1/\sqrt{2})\,d_{z^2})$$
$$= o_1.$$

$o_3$ is obtained from $o_1$ by applying $\hat{C}_2^z$; and $o_6$, $o_4$ and $o_2$ are obtained from $o_5$ by applying $\hat{C}_2^x$, $\hat{C}_4^x$ and $\bar{C}_4^x$, respectively.

## 2.2.  Sulphur hexafluoride, $SF_6$

In the Slater–Pauling description of the bonding in $SF_6$ (or $SX_6$), one electron is placed in each hybrid orbital, $o_i$, and then spin paired with the electron in the ligand orbital to which it is directed. This corresponds to the perfect pairing approximation, as discussed in Chapter 1. However, one of the main drawbacks of this simple theory, which apparently accounts for the existence of $SF_6$, is that no information is forthcoming on the feasibility of the bonding scheme as far as changes in energy are concerned; this is something which can only be settled by detailed calculation. Unfortunately, staggering as it may seem, no proper molecular calculations have yet been made to test the viability of the model—and this more than thirty years after the model was suggested. One of the difficulties is to show that the energy required by sulphur, to enable it to form six bonds, is more than regained on molecule formation— clearly some energy is required because the $3d$ atomic orbitals are not occupied in the ground state of the free sulphur atom (at least not when the ground-state wave function is single configurational in form).

Another problem, which is not treated satisfactorily in the Pauling–Slater model, is concerned with the relative sizes of the valence orbitals used in constructing the hybrid orbitals, $o_i$. These hybrids will only be suitable for bonding with the ligands if the maxima in their respective probability densities lie within the bonding region: that is, the $3s$, $3p$ and $3d$ atomic orbitals of sulphur should all be of similar size. Although this was one of the main assumptions of the Pauling approach, it is difficult to justify, particularly when the $3d$ orbitals are not occupied in the ground state.

All of the deficiencies of the Pauling–Slater method are clearly manifest when seeking a reason for the non-existence of $SH_6$ or $SBr_6$. Any detailed understanding of the factors which determine the stability of compounds involving second-row atoms requires a consideration of the ligand type as well as the electronic properties of the central atom. The realization of the importance and significance of ligand type has led to the application of simple valence bond models (in preference to molecular orbital models); for, it is then possible to separate the intra-atomic contribution to the molecular energy from the terms contributing to the energy of bond formation. This is a useful division of the molecular energy, as it enables the increase in intra-atomic energy (promotion energy) of the second-row atom, arising from the use of the higher energy $3d$ orbitals, to be studied as a function of the ligand type. In this approach, the ground-state wave function for the octahedral $SX_6$ molecule is approximated by

$$\Psi_0 = N\hat{A}\phi_{core}\psi_1\psi_2 \ldots \psi_6, \tag{2.1}$$

where each pair function is of the form

$$\psi_i = (1/\sqrt{2})\{o_i(1)l_i(2) + l_i(1)o_i(2)\}.(1/\sqrt{2})\{\alpha(1)\beta(2) - \beta(1)\alpha(2)\};$$

$l_i$ is an appropriate ligand orbital, and $\phi_{core}$ represents a product of core-pair wave functions associated with the non-valence shell electrons of both sulphur and ligands.

$\Psi_0$ possesses the total symmetry of the $O_h$ point group, since any symmetry operation merely permutes pairs of bond functions, thereby leaving (2.1) invariant. The wave function (2.1) obviously represents an over-simplification of the problem, as it corresponds to only one of the possible singlet totally symmetric covalent pairing schemes involving single occupancy of the $o_i$. In addition, ionic structures and covalent structures, involving doubly occupied hybrid orbitals, $o_i$, are neglected. However, even after making these gross assumptions, which are necessitated by the complexity of the problem, it is hoped to gain insight into the factors which determine the stability of $SX_6$ as a function of X. Substitution of (2.1) into (1.24) then yields (see (1.29))

$$E = \sum_x n_x \int \chi_x \left[ \hat{f} + \tfrac{1}{2} {\sum_y}' n_y(\hat{J}_y - \tfrac{1}{2}\hat{K}_y) \right] \chi_x \, \mathbf{dr}_1 + \tfrac{1}{2} \sum_{x \neq o_i, l_i} n_x J_{xx}$$
$$+ \tfrac{3}{2} \sum_i K_{o_i l_i} \quad (2.2)$$

($x$, $y$ run over the $o_i$, $l_i$ and core orbitals of both sulphur and fluorine ligands) as the expression for the molecular energy (see also (1.29)).

The bonding scheme implied by the use of (2.1) results in a considerable constraint on the sulphur atom: for, in addition to requiring occupancy of the higher energy $3d$ orbitals, each hybrid must be paired with the ligand orbital to which it is directed (the chemically intuitive bonding scheme). In analysing the various contributions to the molecular energy, it is useful to separate out the intra-atomic contribution to $E$, arising from sulphur: this is usually referred to as the valence-state energy, and is given by

$$E_{VS} = 2 \sum_C \int \chi_C \left[ \hat{f} + \tfrac{1}{2} {\sum_{C'}}' (2\hat{J}_{C'} - \hat{K}_{C'}) \right] \chi_C \, \mathbf{dr}_1$$
$$+ \sum_{o_i} \int \chi_{o_i} \left[ \hat{f} + \sum_C (2\hat{J}_C - \hat{K}_C) + \tfrac{1}{2} {\sum_{o_j}}' (\hat{J}_{o_j} - \tfrac{1}{2}\hat{K}_{o_j}) \right] \chi_{o_i} \, \mathbf{dr}_1 \quad (2.3)$$

where $C$, $C'$ label the core orbitals on sulphur. The difference in energy between $E_{VS}$ and the ground-state term, $^4S$, arising from the neutral sulphur valence electron configuration $3s^2 3p^4$, defines the valence state promotion energy.

Although (2.1) is unlikely to be a very good representation of the molecular wave function, (2.3) should at least provide a useful measure of the valence-state energy for comparing the electronic effects of different ligands. This may not be immediately obvious, as (2.3) appears to be independent of the nature of the ligands. However, the Coulomb and exchange interactions between sulphur and ligands determine the optimum sizes of the sulphur valence orbitals; and so (2.3) will vary according to the exponents required to optimize $E$ for a given ligand.

A closer examination of (2.3) shows that after expanding the hybrid atomic orbitals, in terms of their constituent atomic orbitals—here $3s$, $3p$, and $3d$—a rather complicated expression results which involves multiple sums of integrals over ordinary atomic orbitals. In fact, this new form of $E_{VS}$ represents a weighted sum of the energies of selected terms arising from the sulphur electronic configurations $3s^2 3p^4$, $3s3p^5$, $3p^6$, $3s3p^3 3d^2$, ... (maximum $d$-orbital occupancy for any configuration is four). To see how this arises, it is instructive to work through the analysis for an atom forming only two bonds; as, for example, with $B^+$ in linear $BH_2^+$, which was one of the molecules discussed in Chapter 1.

The valence-state energy of $B^+$, in the perfect pairing approximation, and neglecting the energy of the core electrons, is given by

$$E_{VS} = 2I_{BB} + J_{BB} - \tfrac{1}{2}K_{BB},$$

where $I_{BB}$ represents the sum of the average kinetic energy, nuclear potential energy, and the Coulomb and exchange interactions of an electron in an hybrid orbital with the $1s$ core electrons. $J_{BB}$ and $K_{BB}$ are the usual inter-hybrid Coulomb and exchange energies, respectively. On expanding the hybrids according to (1.16), the expression for the valence-state energy becomes

$$I_{ss} + I_{pp} + \tfrac{1}{4}J_{ss} + \tfrac{1}{4}J_{pp} - K_{sp} + \tfrac{1}{2}J_{sp} - \tfrac{1}{8}J_{ss} - \tfrac{1}{8}J_{pp} + \tfrac{1}{4}J_{sp},$$

where integrals like $I_{sp}$ and $(ss/sp)$ disappear for symmetry reasons (the integrand is odd with respect to inversion through the origin of coordinates). Thus, on collecting terms, $E_{VS}$ becomes

$$I_{ss} + I_{pp} + \tfrac{1}{8}J_{ss} + \tfrac{1}{8}J_{pp} - K_{sp} + \tfrac{3}{4}J_{sp}$$

which is, in fact, a weighted sum of term energies associated with the electron configurations $2s^2$, $2s2p$, and $2p^2$ of $B^+$. The energy expressions for all the terms arising from these configurations are obtained from the term wave functions as given by Slater:[4]

$$E(2s^2, {}^1S) = 2I_{ss} + J_{ss},$$

$$E(2p^2, {}^3P) = 2I_{pp} + J_{pp'} - K_{pp'},$$

$$E(2p^2, {}^1D) = 2I_{pp} + J_{pp} - K_{pp'},$$

$$E(2p^2, {}^1S) = 2I_{pp} + J_{pp} + 2K_{pp'},$$

$$E(2s2p, {}^3P) = I_{ss} + I_{pp} + J_{sp} - K_{sp};$$

$$E(2s2p, {}^1P) = I_{ss} + I_{pp} + J_{sp} + K_{sp},$$

where $p$, $p'$ label different $p$-orbitals in real form (notice that Slater[4] uses complex $p$-orbitals, with a phase convention given in his equation (9.1) of Chapter 9, volume 1—see also p. 116).

Since $J_{sp}$ and $K_{sp}$ arise only from the energies of $2s2p$, ${}^3P$ and $2s2p$, ${}^1P$, the weights of these term energies, say $x$ and $y$ respectively, are obtained from the solution of the two equations $x + y = \frac{3}{4}$, and $-x + y = -1$; that is, $x = \frac{7}{8}$ and $y = -\frac{1}{8}$. And as $J_{ss}$ only appears in the expression for $E(2s^2, {}^1S)$, the weight of this term in $E_{VS}$ is $\frac{1}{8}$. Hence,

$$E_{VS} = \tfrac{1}{8}E(2s^2, {}^1S) + aE(2p^2, {}^3P) + bE(2p^2, {}^1D) + cE(2p^2, {}^1S)$$
$$+ \tfrac{7}{8}E(2s2p, {}^3P) - \tfrac{1}{8}E(2s2p, {}^1P).$$

However, $a = 0$ because $J_{pp'}$ appears in only one term energy, and does not appear in $E_{VS}$. Also $b$ must equal $2c$, in order to eliminate terms in $K_{pp'}$. Finally, the correct weight of $J_{pp}$ is obtained only by having $b = \frac{1}{12}$ and $c = \frac{1}{24}$. The resulting expression for the valence state energy:

$$E_{VS} = \tfrac{1}{8}E(2s^2, {}^1S) + \tfrac{1}{12}E(2p^2, {}^1D) + \tfrac{1}{24}E(2p^2, {}^1S) + \tfrac{7}{8}E(2s2p, {}^3P)$$
$$- \tfrac{1}{8}E(2s2p, {}^1P),$$

therefore shows (a) how the central atom spin and orbital coupling schemes have been broken, and (b) that the promotion energy may be large because of the small weighting of the lowest energy term of $B^+$. In fact, the calculations of Hinze and Jaffé[5] show a promotion energy of 0·75 aJ. It should also be noted that as $\hat{H}$ is not diagonal with respect to the term wave functions listed above (there is a matrix element between the two $^1S$ terms), it is not strictly possible to utilize experimental term energies in estimating $E_{VS}$. However, the effects of configuration interaction are often neglected, as the zeroth order states are usually well separated in energy (see Hinze and Jaffé[5]).

A similar calculation for sulphur, using the six hybrid orbitals, $o_i$, is more involved, but the principle remains the same. Fortunately, the analysis can be simplified by the use of powerful group theoretical techniques; and the reader is referred to the work of Craig and Thirunamachandran[6] for further details, and for particulars of the earlier literature. In this particular instance, the magnitude of the problem can be appreciated when it is realized that the valence-state energy, within the perfect pairing approximation, has contributions from nineteen terms arising from the $sp^3d^2$ configuration of neutral sulphur, as well as terms from the other possible valence electron configurations.

Preliminary investigations into the plausibility of the octahedral hybrid model for sulphur compounds have been made by several authors. Some authors[7] prefer to focus attention on the calculation of $E_{VS}$, while others[8-9] prefer a direct, but approximate, calculation of $E$. The main problem in either approach lies in the particular choice of sulphur valence orbitals. There is no problem with the $3s$ and $3p$ atomic orbitals, as their size, as measured by the outermost maximum in their respective radial distribution functions (for Slater orbitals this occurs at $3/k$ a.u., where $k$ is the orbital exponent), is not very sensitive to the electron configuration: for example, the sizes of the $3s$ and $3p$ functions are very similar in the terms arising from the $3s^23p^4$ and

$3s3p^33d^2$ electron configurations of sulphur (see Table 2.1). The same situation does not hold for $3d$ atomic orbitals. Atomic Hartree–Fock calculations on selected excited states of sulphur, show that even within the same electron configuration, different terms may involve $d$ orbitals of significantly different size. The $d$-orbital size also varies with $d$-electron configuration—again, in complete contrast to the case of $3s$ and $3p$ atomic orbitals. All of these effects are clearly seen in the results of Coulson and Gianturco,[7] which are summarized in Table 2.1.

TABLE 2.1

*The positions of the maxima in the radial distribution functions of the valence atomic orbitals of neutral sulphur, as determined from the atomic Hartree–Fock calculations of Coulson and Gianturco[7] ($r_m$ is the value of r for which $[rg_d(r)]^2$ attains its maximum value).*

| Configuration | State | $r_m(3s)$ | $r_m(3p)$ | $r_m(3d)$ |
|---|---|---|---|---|
| $3s3p^33d^2$ | $^3I$ | $0 \cdot 074$ nm | $0 \cdot 079$ nm | $0 \cdot 142$ nm |
| | $^5H$ | $0 \cdot 074$ | $0 \cdot 079$ | $0 \cdot 130$ |
| | $^7F$ | $0 \cdot 074$ | $0 \cdot 079$ | $0 \cdot 120$ |
| $3s^23p^33d$ | $^5D$ | $0 \cdot 074$ | $0 \cdot 085$ | $0 \cdot 334$ |
| $3s^23p^4$ | $^3P$ | $0 \cdot 074$ | $0 \cdot 082$ | — |
| $0_10_20_30_40_50_6$ | perfect pairing valence state | $0 \cdot 074$ | $0 \cdot 079$ | $0 \cdot 106$ |

The variation of $d$-orbital size with atomic configuration makes it imperative to select the most suitable $3d$ radial function for the molecular calculation. A particular choice of free atom function is unlikely to be realistic—even if it has been calculated by the atomic Hartree–Fock procedure—as the ligands will always cause some distortion of the electron density in the vicinity of the sulphur atom. These distortions can involve changes in shape and size; but changes in size, as evidenced by a change in the maximum of the radial distribution function, are usually considered more important.

This problem was recognized some years ago by Craig and Zauli,[8] who showed that the free atom sulphur $3d$ orbitals are very diffuse when described by simple Slater orbitals, with exponent chosen by Slater's rules. But they found that six octahedrally disposed fluorine atoms, taken as sources of a perturbing potential, caused a considerable contraction of the $3d$ orbitals, thereby making them suitable for combining with $3s$ and $3p$ orbitals in the construction of hybrid orbitals. Although it was found later (see Table 2.1) that the $3d$ orbitals are always relatively compact, when determined as solutions of the atomic Hartree–Fock equations, it is still crucial to optimize the size of the $3d$ orbital in the molecular environment. It is highly unlikely, for example, that the very compact $3d$ orbitals found by Coulson and

Gianturco[7] (see last entry in Table 2.1), in the optimization of $E_{VS}$, will persist in the molecular environment. For this reason, it is convenient to work with a $3d$ orbital which has some kind of built-in flexibility. One possible approach is to fit a number of Slater $3d$ functions to a suitable atomic Hartree–Fock orbital in numerical form. A uniform scaling factor, $c$, can then be introduced to allow the orbital to contract or expand, depending upon the value of $c$ required to minimize the molecular energy:

$$g_d' = c^{3/2} \sum_n a_n g_d(k_n cr)$$

where

$$g_d(k_n cr) = [(2k_n)^7/6!]^{1/2}(cr)^2 \, e^{-k_n cr}.$$

The problem of calculating $E$ is immense and, apart from the difficulty in selecting the appropriate radial form for the $3d$ orbitals, there still remains the problem of allowing for the real effects of non-orthogonality between the sulphur valence orbitals and the ligand orbitals: particularly the ligand core orbitals. The neglect of this non-orthogonality allows the sulphur valence electrons to penetrate the ligand electron density far too readily; thus, important repulsive contributions are missing from $E$ which, for heavy ligands, such as Br, may more than offset the energy gained in the ligand field. Tentative calculations by Mitchell[9] suggest that this may be one of the factors contributing towards the non-existence of $SBr_6$ as against $SF_6$. Estimates for the magnitude of this effect, which certainly cannot be neglected, differ significantly; the main problem is to calculate its contribution to $E$ in a reliable way, but this is not possible at the present time. As seen already, though, the work of Stuart and Hurst on LiH shows very clearly the danger in neglecting the effects of valence-core non-orthogonality. In this instance, the failure to orthogonalize $1s_H$ to $1s_{Li}$ resulted in a molecular energy which is lower than the estimated "experimental" value. The inclusion of the neglected repulsive contributions to the energy ameliorates the situation completely. Thus, even in a molecule as simple as LiH, the non-orthogonality corrections play an important role, and should act as a warning when interpreting the results of calculations on $SX_6$ molecules.

The above discussion on the ability of sulphur to utilize octahedral hybrids has centred mainly on the problem of orbital size. But this is only one of the factors contributing towards the formation of strong bonds; the difference in the appropriate orbital energies of sulphur and ligand must also be small. The results of Hartree–Fock calculations on terms arising from the $3s3p^33d^2$ electron configuration of sulphur, which produce well-contracted orbitals, show that the electrons in the $d$-orbitals are weakly bound. But Craig and Zauli,[8] and, more recently, Cruickshank, Webster and Spinnler[10] have shown that an octahedral disposition of fluorine atoms can result in $d$-orbital stabilization in the molecule. A further interesting point arises from

Cruickshank, Webster and Spinnler's calculations on selected terms of $S^+$ electron configurations: the $3d$ orbitals are found to be strongly bound, and well contracted, and these are both requirements for good bonding. It therefore appears that ligands which can accept charge from sulphur will favour participation by sulphur $3d$ atomic orbitals. This observation suggests that the ground state may be more adequately described in terms of an ionic model, than by the covalent perfect pairing model; a result which is also supported by some calculations of Craig and Zauli[8] on various ionic states of sulphur in octahedral fields of fluorine ligands. However, their conclusions must be regarded with some care, as resonance between equivalent structures was not permitted—nor were the problems of orbital non-orthogonality satisfactorily treated.

The results of the exploratory calculations by Craig and Zauli,[8] Mitchell,[9] and other authors suggest a large promotion energy of about 4·8 aJ for sulphur, which is expected to be more than offset by the energy released on bond formation—particularly if the ligands are fluorine atoms. The de-stabilizing influence of the repulsive valence-ligand core penetration energies still remains uncertain, although there have been some estimates of the magnitude of the effect.[9-11]

If $d$-orbital participation is important in compounds like $SF_6$, it seems probable that the ability of fluorine to stabilize $d$-orbitals, in the molecular environment, arises partly from its ability to polarize the distribution of electrons around sulphur, and partly from its ability to accept charge from sulphur. Both of these factors greatly favour fluorine over hydrogen as a ligand. However, an interesting situation now arises, for if the structure of $SF_6$, and other high valence states of second-row elements, is determined more by the possibility of delocalizing charge from the central atom to the ligands, then the general stereochemical features of these compounds can be understood without recourse to the use of $d$-orbitals. Although the use of $d$-orbitals is still often taken for granted, it is very difficult to justify the large amount of $d$-orbital character which is required in the construction of the octahedral hybrids. Fortunately, a valence bond description of the bonding can be found which does not depend fundamentally on the participation of $d$-orbitals: their use is regarded more in terms of giving the finer details of the changes in shape of the electronic charge distribution, rather than forming an integral part of the theory (in molecular orbital theory there are also no problems with $d$-orbitals, as described shortly).

In this approach, which is an extension of Pauling's model, the bonding is assumed to involve only the valence $3s$ and $3p$ atomic orbitals of sulphur. This limits the maximum number of electron pair bonds to four, so two elec-trons must be first transferred from sulphur to the ligands: in fact, one to each of two fluorine atoms. For ease of description, two $sp$ hybrid orbitals are then constructed for use by $S^{++}$ (the promotion energy is expected to be small), while the other sulphur valence orbitals involved in the bonding are the two remaining pure $3p$ atomic orbitals. A plausible pairing scheme is shown in Fig. 2.2. In a more complete approach, of course, it is necessary to consider alternative coupling schemes and hybrid orbital occupancies; but the chemically intuitive coupling scheme is assumed dominant for the purposes of the present discussion.

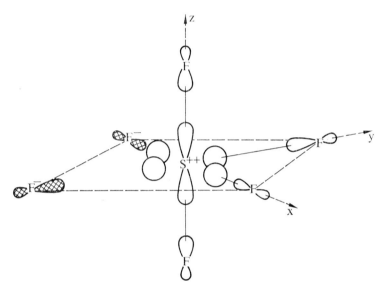

F₁ɢ. 2.2. A typical valence bond structure for $SF_6$ based on $S^{++}$. The orbitals used by sulphur for bonding are the $p_x$, $p_y$ and $sp_z$ hybrid atomic orbitals. Interatomic electron pairing is indicated by a line joining singly occupied orbitals, and orbitals containing lone-pairs of electrons are indicated by crossed hatching.

There are three more structures of the kind shown in Fig. 2.2, which differ only in the positioning of the two (*cis*) F⁻ ions. Also, two further choices of the $p$-atomic orbital used in constructing the $sp$ hybrids, give rise to another eight structures. There are therefore twelve structures in all, and it is possible to construct one linear combination of structures transforming like the totally symmetric irreducible representation of $O_h$. The lowest energy structure is expected to involve the in-phase combination of the twelve structures with equal weights, so leading to a regular octahedral structure with a formal charge of a third of an electron residing on each ligand. The existence of very low-lying excited states, thereby allowing for the possibility of molecular distortions, arising through the effects of vibronic coupling, is considered unlikely in the present situation because the matrix elements between structures are expected to be large in magnitude. The energy required to form the structure given in Fig. 2.2 is certainly less than the second ionization potential of sulphur, because of the favourable gain in electrostatic energy on separating the charges. A very rough estimate shows that although 5·45 aJ are required to doubly ionize sulphur, a further 3·84 aJ are gained after allowing for the electrostatic interactions between the separated formal charges, using an S—F bond length of 0·15 nm. The net energy required is therefore only 1·60 aJ: an amount which should be regained easily on bond formation.

This model, which, by necessity, predicts a charge displacement from sulphur to ligands, is expected to lead to a contraction of the sulphur valence atomic orbitals, thereby making them more suitable for forming strong bonds. An interesting feature of the model is concerned with the repulsive energy arising from the penetration of the two $F^-$ ions into the $3p$ electron density of sulphur. This repulsive energy is expected to be smallest for fluorine, on account of size, and to be greatly increased for a ligand, like bromine, with more core electrons. Also, since hydrogen is a poor electron acceptor, and $H^-$ has a rather expansive distribution of electron density, the ability of fluorine to stabilize octahedral sulphur can be understood in terms of its size and high electronegativity.

Very similar conclusions are obtained from a simple molecular orbital model in which $d$-orbitals are excluded, as originally discussed by Rundle.[12] As described in Chapter 1, the molecular orbitals are delocalized over the molecular framework, and can be characterized by the symmetry of the nuclear framework (at least as far as closed shells are concerned, as $\hat{h}$ commutes with all symmetry operations of the group). The molecular orbitals are best constructed in terms of linear combinations of symmetry orbitals in the form

$$\psi_j(\Gamma) = \sum_M d_{Mj}(\Gamma)\lambda_M(\Gamma) + \sum_L d_{Lj}(\Gamma)\lambda_L(\Gamma)$$

where $\lambda_M$ and $\lambda_L$ are suitable combinations of central atom and ligand atomic orbitals, respectively, transforming like the irreducible representation $\Gamma$ of the molecular point group $G$. The symmetry orbitals themselves (not normalized) are generated by standard group theoretical techniques, as described in Vol. 1:

$$\lambda_L(\Gamma) = \left[\sum_{\hat{R}} \chi_\Gamma^*(\hat{R})\hat{R}\right]\phi_\alpha$$

where $\phi_\alpha$ is an arbitrary ligand atomic orbital. The operator in square brackets is applied to different ligand atomic orbitals until the required number of linearly independent symmetry orbitals is found: in the case of a degenerate irreducible representation, it may be necessary to orthogonalize the resulting symmetry orbitals before proceeding with the molecular orbital calculation.

The particular combinations of atomic orbitals required for $T_d$ and $O_h$ are shown in Appendix III. The combining coefficients of the various symmetry orbitals are determined by the analysis given in Chapter 1, but the basic one-electron functions are now multicentre symmetry orbitals rather than mono-centric atomic orbitals.

It is a major problem to perform accurate molecular orbital calculations on molecules as complex as $SF_6$, but the general pattern of the results can often be obtained by qualitative reasoning. In $SF_6$, for example, where only $3s$, $3p$ atomic orbitals on sulphur and $2p_\sigma$ atomic orbitals on fluorine are considered, the ten atomic orbitals give rise to ten molecular orbitals in the usual way.

Application of the group theoretical techniques, discussed in the earlier part of this Chapter, shows that the reducible representation, which is generated by the set of $10 \times 10$ matrices, breaks down into the irreducible representations

$$2a_{1g} + 2t_{1u} + e_g.$$

The $e_g$ molecular orbital is constructed entirely out of ligand atomic orbitals, as the sulphur $3s$ and $3p$ atomic orbitals span the irreducible representations $a_{1g}$ and $t_{1u}$, respectively: the $e_g$ molecular orbital is usually described as non-bonding. The remaining molecular orbitals occur in pairs of the same symmetry, corresponding to bonding and antibonding (virtual) molecular orbitals.

The twelve electrons are therefore accommodated in the six lowest energy molecular orbitals, giving a ground state electron configuration $a_{1g}^2 t_{1u}^6 e_g^4$—this is the main result, because the actual ordering of the orbital energies, although important for correlating with observed ionization potentials, is not vital here. A plausible pattern of the occupied molecular orbital energy levels is given by

but it is quite possible that actual calculations may lead to a different ordering of the levels. In fact, the calculation made by Brown and Peel,[13] using the same set of atomic orbitals, leads to the pattern shown above.

If the reasonable assumption is made, whereby the electrons in the $a_{1g}$ and $t_{1u}$ molecular orbitals are shared equally between sulphur and the ligands, then sulphur has effectively lost two electrons to the ligands, as the remaining four electrons, in the non-bonding $e_g$ molecular orbital, only contribute to the electron density on each ligand. Thus, both the simple molecular orbital and valence bond theories, in their forms which exclude participation by $d$-orbitals, lead to a description of the charge distribution in terms of a formal charge transfer from sulphur to the ligands. It should be remembered, of course, that the notion of charge transfer is difficult to quantify; and even in the covalent structure, based on the use of six singly occupied hybrid orbitals, $o_i$, it could be argued that charge is transferred away from the vicinity of the sulphur atom. This arises because the use of hybrid orbitals, which are concentrated

D

mainly in the region of the ligands, leads to bond densities with centroids closer to the ligands than to sulphur. Hence, although the discussion of the bonding in terms of charge transfer processes is necessarily subjective, it still provides a useful scheme for comparing the electronic effects of different ligands.

A very interesting aspect of the molecular orbital treatment becomes apparent if the molecular orbitals are subjected to a unitary (or orthogonal) transformation to obtain the equivalent, or localized, molecular orbitals (see Vol. 1 for a discussion of molecular orbital-equivalent orbital transformations). The group theoretical discussion given in the earlier part of this chapter, showed that a set of six geometrically equivalent hybrid orbitals can be constructed out of atomic orbitals having $a_{1g}$, $t_{1u}$, and $e_g$ symmetries in the octahedral environment. But the symmetry argument is not basically concerned with the nature of the one-electron functions, and so if these are taken as molecular orbitals, the same transformation will lead to a set of geometrically equivalent localized, molecular orbitals: the only prerequisite being the availability of molecular orbitals with symmetry $a_{1g}$, $t_{1u}$ and $e_g$. This requirement is clearly satisfied. The equivalent molecular orbitals will not be localized entirely within the region of a chosen sulphur–fluorine bond, as the LCAO coefficients of other fluorine atomic orbitals cannot be reduced to zero by the transformation. Nevertheless, it makes a reasonable approximation to envisage each equivalent molecular orbital as encompassing just two nuclei.

It is now possible to envisage the electronic structure of $SF_6$ in terms of six doubly occupied localized one-electron functions. Although the description of the bonding now appears to be different from the one provided by molecular orbital theory, the nature of the transformation to localized orbitals ensures that the calculated physical properties, such as energy, dipole moment, electron density function, etc., remain unaltered (see Section 1.5 for a proof of the invariance of $E$ under orthogonal transformations).

The pertinent point of the molecular orbital approach becomes evident, when it is noticed that there are sufficient occupied molecular orbitals of the appropriate symmetry to enable the transformation to equivalent molecular orbitals to be carried out. This is where the Pauling–Slater and simple valence bond approaches differ; the Pauling–Slater analysis is based solely on the preparation of the central sulphur atom to form six geometrically equivalent bonds. This requirement necessarily involves the introduction of the higher energy $3d$ orbitals in order to make the transformation to localized atomic orbitals feasible. On the other hand, with the simple molecular orbital approach, the $e_g$ orbitals required for the transformation to localized functions are provided by the *ligand* atomic orbitals. The simple valence bond approach, described above, which avoids the use of $d$-orbitals, produces bond equivalence by imposing the requirements of octahedral symmetry on the possible choice of structures. Thus, the weakness of the Pauling–Slater approach comes more from the rigidity of thought which requires the central sulphur atom to be prepared in the state of maximum covalency—a feature which London, and Heitler and Rumer, developed in the early days of valence theory, and which the molecular orbital theory, in particular, shows is unnecessary.

Both the simple molecular orbital and valence bond methods described in this chapter can be improved by addition of $3d$ orbitals to the basic set of atomic orbitals. The degree of $d$-orbital participation can then be determined variationally, in an attempt to improve the description of the change in shape of the electronic charge distribution, around sulphur, as a result of molecule formation. The effects of $d$-orbital participation in both models are now considered.

The addition of $d$ orbitals in the simple valence bond scheme leads to the existence of many more structures. In fact structures can now be formulated which involve $S^+$, rather than $S^{++}$: a typical example of which is shown in Fig. 2.3 where, for simplicity, diagonal hybrids are used by sulphur for bonding to two of the fluorine ligands, and the sulphur $3s$ orbital accommodates a lone pair of electrons. As before, there are twelve structures of this kind, giving rise to one $^1A_{1g}$ molecular wave function. The equivalence of $d_{z^2}$ and $d_{x^2-y^2}$, under $O_h$, dictates that the corresponding set of structures involving $d_{x^2-y^2}$ must also be considered.

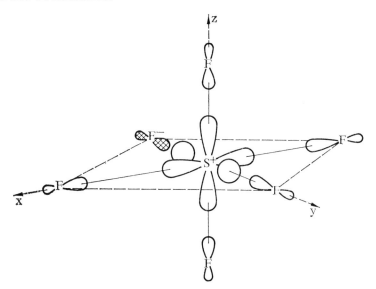

Fig. 2.3. A typical valence bond structure for $SF_6$ based on $S^+$. The orbitals used by sulphur for bonding are the $p_y$ atomic, and $sp_x$ and $p_z d_{z^2}$ hybrid atomic orbitals. Interatomic electron pairing is indicated by a line joining singly occupied orbitals, and the orbital containing the lone-pair is indicated by crossed hatching.

Although there are several other ways of effecting suitable pairing schemes, the main point about this approach, which is missing in the Pauling–Slater method, is that the amount of $d$-orbital participation is determined variationally. For example, if some of the pairing schemes are unfavourable, this will be apparent from their small contribution to the molecular wave function.

The valence-state energy of sulphur now has contributions from terms arising from configurations involving $3s$, $3p$, and $3d$ orbitals of $S^+$ (and $S^{++}$); but the weights of terms involving $d$-orbitals are determined variationally and, hence, their contribution will depend more directly on the nature of the ligand. This is in complete contrast to the perfect pairing model, in which the contribution of the various terms to the valence-state energy is fixed by the choice of hybrid orbitals (in this particular instance the hybrid orbitals are determined completely by the molecular symmetry).

The analogous extension of the simple molecular orbital theory is made by adding sulphur $3d$ orbitals to the set of atomic orbitals used in expanding the molecular orbitals. As in the extension of the valence-bond method, it is sufficient to add only sulphur $d_{z^2}$ and $d_{x^2-y^2}$ orbitals to the basis set, so long as $\pi$-bonding is neglected. An interesting situation now arises, though, because the $3d$ orbitals on sulphur, and an appropriate combination of ligand atomic orbitals, both possess $e_g$ symmetry. The previously non-bonding $e_g$ molecular orbital, concentrated in the region of the ligands, is now stabilized through the interaction with the sulphur $e_g$ orbitals, thereby leading to a shift of the electron density from the region of the ligands towards the sulphur atom. Consequently, the inclusion of sulphur $d$-orbitals reduces the extent of formal charge transfer which is required in the simple model. Interestingly enough, this conclusion exactly parallels the one reached by extending the valence-bond approach: for here it was found that the inclusion of $d$-orbitals enabled structures to be fabricated which had a formal charge of $+1$ on the sulphur atom, as against $+2$ in the simple approach.

The stability of certain hexavalent sulphur compounds now becomes more easily understood. First, an electron-accepting ligand is required (high $\sigma$-orbital electronegativity) to withdraw electron density from around the sulphur atom, in order to make the $d$-orbitals more accessible. The stabilized $d$-orbitals are then of a suitable size, and energy, to permit a back-transfer of electron density from the ligands to the region of the sulphur atom, leading to a further stabilization of the molecule.

In so far as effective atomic charges give useful information about the molecular electronic charge distribution, recent semi-empirical molecular orbital calculations by Brown and Peel,[13] Hillier,[14] and Santry and Segal[15] on $SF_6$ show that the effective atomic charge on sulphur is reduced from about $+2\cdot2$ to $+1\cdot1$ by the inclusion of $d$-orbitals; apparently supporting the conclusions reached above. However, as the maximum in the $3d$ radial distribution function lies nearer to fluorine than sulphur, particularly with

Hillier's choice of $d$-orbital exponent, it may be improper to ascribe charge density in sulphur $d$-orbitals as pertaining to sulphur; also, the procedure of equipartitioning the overlap densities, in the calculation of the effective atomic charges, is clearly suspect. A much more meaningful result would be obtained from the plot of the $\Delta P$ function, for calculations with and without $d$-orbitals; but, unfortunately, the use of empirical values for many of the integrals contributing to the matrix elements of $\hat{h}$ precludes the possibility of any reliable density maps being drawn (the basic atomic orbitals are not unambiguously defined).

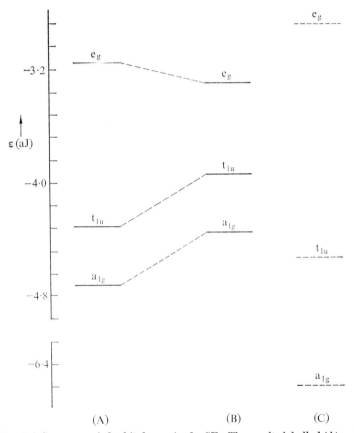

FIG. 2.4. Some occupied orbital energies for SF$_6$. The results labelled (A) and (B) were obtained by Hillier[14] in calculations with and without participation by $d$-orbitals, respectively. In calculation (A) the basis set included the $3s$, $3p$ atomic orbitals of sulphur and the $2s$, $2p_\sigma$ and $2p_\pi$ atomic orbitals of the fluorine ligands. The results shown in column (C) were obtained by Brown and Peel[13] in calculations using only $3s$, $3p$ atomic orbitals on sulphur and the $2p_\sigma$ atomic orbitals of the fluorine ligands. The (A) and (C) sequences of levels, although similar, differ in detail because Hillier was able to treat the polarization of the $\pi$-electron density in a self-consistent manner.

The change in the pattern of the molecular orbital energies, on including participation by $d$-orbitals, illustrates the point made earlier, where the occupied $e_g$ molecular orbital now becomes more strongly bound (see Fig. 2.4).

The simple valence-bond model, discussed above for $SF_6$, is readily extended to other compounds involving a central second-row atom. To illustrate the simplicity of the approach, the remainder of this chapter is devoted to the problem of examining plausible geometries for $SF_4$, $SF_5$, $PF_5$ and $ClF_3$: but it must be emphasized again that the model provides only a qualitative description of the bonding—a more quantitative approach will have to await the results of further calculations.

## 2.3.  Sulphur tetrafluoride, $SF_4$

$SF_4$ contains five pairs of valence electrons, and the most likely stereochemistries are therefore square pyramidal or trigonal bipyramidal; or at least some geometry close to one of these basic types. The square pyramidal configuration can involve the lone pair in either an equatorial (A), or axial position, (B) or (C) (see Fig. 2.5). In (A) one of digonal hybrids in the plane accommodates a lone pair, while the other hybrid is used for bonding with a fluorine atom. The two other fluorine atoms are bonded through the sulphur $p$-atomic orbitals. There are obviously several other equivalent structures of type (A) which contribute to the molecular wave function, all of which are associated with a formal charge transfer of one electron from sulphur to fluorine.

The situation where the four fluorine atoms are in the same plane leads to two sets of structures which, in a rigorous approach, must be considered together. However, in this qualitative approach, it is sufficient to regard the structures (B) and (C) as distinct forms; (C) is clearly energetically unfavourable in relation to (A) or (B). And (B) is expected to be less favoured than (A), because the lone-pair, being in a pure $p$-orbital, is in close proximity to three S–F bond pairs rather than two, as in (A). Hence, structure (A) is favoured on grounds of intuitive reasoning, and leads to the expectation that there will be two kinds of F–S–F linkage: one linear and the other bent, in broad agreement with the experimental data. The trigonal bipyramidal structure (D) also yields linear and bent F–S–F linkages, but the two equatorial fluorine atoms are now at an angle of 120° to each other (see Fig. 2.6). Structure (D) also requires a formal transfer of one electron from sulphur to fluorine, just as in the square pyramidal model, and the only real difference between the two structures is in the amount of $s$ character in the hybrids. As this must be determined variationally, the structure with the minimum energy is expected to lie somewhere between the limits of (A) and (D): that is, the bent F–S–F linkage is expected to be characterized by an angle somewhere between 90°

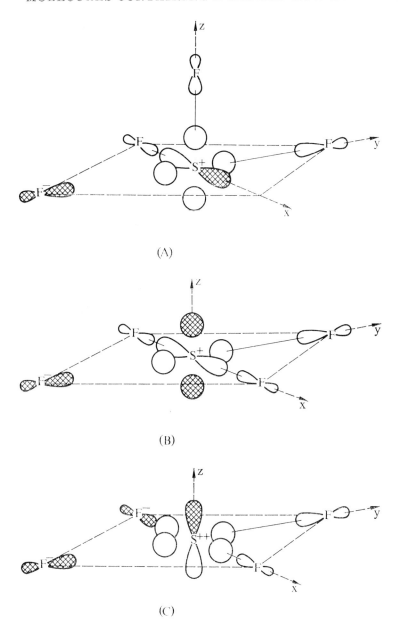

(A)

(B)

(C)

Fig. 2.5. Typical valence bond structures for SF₄. In A and B, the $sp_x$ hybrid and $p_y$, $p_z$ atomic orbitals are used by sulphur for bonding with the ligands. Interatomic electron pairing is indicated by a line joining singly occupied orbitals, and orbitals containing lone-pairs of electrons are indicated by crossed hatching. Structure C is based on the use of sulphur $sp_z$ hybrid and $p_x$, $p_y$ atomic orbitals.

and 120°. In fact an angle nearer to 90° is expected, as the formation of $sp$ hybrids requires less promotion energy for $S^+$, than does the formation of $sp^2$ hybrids: a result which follows because structure (D) is expected to give greater weight to structures containing the high energy $sp^4$ and $p^5$ configurations of $S^+$.

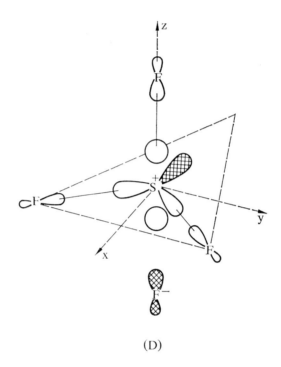

(D)

Fig. 2.6. A typical valence bond structure for $SF_4$ based on the trigonal bipyramidal arrangement of ligands around $S^+$. $sp^2$ $(p_x, p_y)$ hybrid and $p_z$ atomic orbitals are used by sulphur. Interatomic electron pairing is indicated by a line between singly occupied orbitals, and orbitals containing lone-pairs of electrons are indicated by crossed hatching.

The role of $d$-orbitals must also be investigated, in common with the treatment of $SF_6$, as their inclusion permits structures to be formulated which are based on the states of neutral sulphur. In this respect, it is interesting to note that the preliminary molecular orbital calculations of Brown and Peel,[16] at the experimentally observed molecular geometry, show only a small $d$-orbital contribution to the molecular wave function.

Willett[17] has also discussed the electronic structure of $SF_4$, without resorting to the use of $d$-orbitals; but his analysis is also based on the assumption of the experimentally observed molecular geometry.

## 2.4.    Sulphur pentafluoride, $SF_5$

The geometry expected for the $SF_5$ radical is of some interest, even though it has not yet been isolated, because it is thought to be an intermediate in several reactions involving related molecules.[18] For example, $S_2F_{10}$ is produced on irradiating $SF_5Cl$ with ultra-violet light; also, mechanistic studies of the addition of $SF_5Cl$ to olefins suggest that the reaction proceeds through the initial formation of $SF_5$ radicals.

On the basis of the simple valence bond model, developed above, the geometry of $SF_5$ follows directly from that of $SF_6$. One of the axial fluorine atoms in Fig. 2.5(A) is removed, leaving an unpaired electron in the $sp$ hybrid perpendicular to the plane containing the four fluorine ligands (see Fig. 2.7).

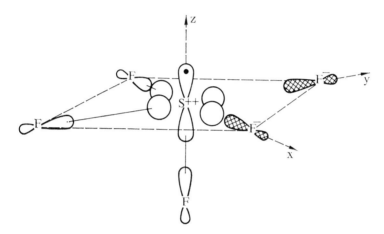

Fig. 2.7. A typical valence bond structure for $SF_5$. The orbitals used by sulphur for bonding are $p_x$, $p_y$ atomic and $sp_z$ hybrid atomic orbitals. Interatomic electron pairing is indicated by a line joining singly occupied orbitals, and orbitals containing lone-pairs of electrons are indicated by crossed hatching.

The other possibility of removing one of the fluorine ligands bonded to a pure sulphur $3p$ atomic orbital is not favoured, as this leads to a structure in which the $F^-$ ion is penetrating the orbital containing the unpaired electron (although this structure will contribute to the molecular wave function, its weight is expected to be small).

Thus, the $SF_5$ radical is expected to be square pyramidal. Furthermore, with the unpaired electron in the hybrid orbital, perpendicular to the plane containing the four fluorine ligands, it is perhaps not surprising that two square pyramidal $SF_5$ units pair up to form the stable dimer $S_2F_{10}$.

## 2.5.    Phosphorus pentafluoride, $PF_5$

The electronic structure of $PF_5$ is usually discussed in terms of the Pauling–Slater model, which requires the use by phosphorus of $sp^3d$ bipyramidal hybrids. Clearly, it is impossible to ascertain the feasibility of this scheme, without a detailed calculation of the promotion energy required in attaining the appropriate valence state. The simple valence-bond approach, adopted here, which initially excludes participation by $3d$ orbitals, suggests one of two possible geometries for the molecule; these are schematically indicated by the representative square pyramidal (E) and trigonal bipyramidal (F) structures shown in Fig. 2.8.

A very interesting situation now obtains, because there is no obvious criterion for selecting the best structure: in fact, both structures are expected to have very similar energies. This conclusion is given some support by the results of Hinze and Jaffé,[5] who find that the promotion energies for $P^+$, in the valence states used in (E) and (F), are 1·18 aJ and 1·14 aJ, respectively. The energy released in forming four electron pair bonds, and one ionic bond, is also expected to be similar in both cases. Berry and his co-workers[19] also find that twice the sum of the occupied molecular orbital energies for the square pyramidal and for the trigonal bipyramidal forms of $PF_5$ differ by only 0·46 aJ. Although there are many imponderables when using the sum of the orbital energies as a measure of the total energy, the fact that the two sums are very similar ($-167·7$ aJ and $-168·2$ aJ, respectively) may well give an indication of a similarity in total energies.

The experimental evidence on the structure of $PF_5$ indicates that the fluorine atoms undergo vibrational motion of large amplitude;[20-22] in fact, under appropriate conditions the square pyramidal and trigonal bipyramidal forms are interconvertible. It is therefore not surprising that some difficulty is experienced in trying to find a unique geometry with which to describe the molecule.

The expansion of the atomic orbital set, to include $d$ orbitals on phosphorus, proceeds in a similar way to that used in $SF_6$: structures involving $sp^3d$ configurations of phosphorus can be constructed, which help in decreasing the amount of charge separation in the molecule. But, to avoid spurious results, it is important to optimize the size of the $d$-orbital in the molecular environment, as the $sp^3d$ configuration of the free phosphorus atom is characterized by a diffuse $3d$ orbital.

## 2.6.    Chlorine trifluoride, $ClF_3$

In $ClF_3$ there are five electron pairs to be accommodated and, once again, this suggests the possibility of a square pyramidal, or trigonal bipyramidal,

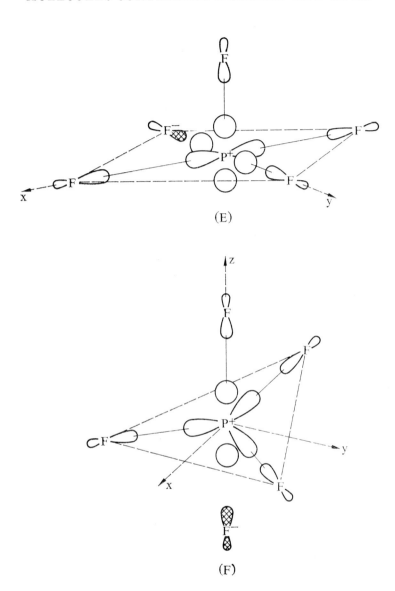

(E)

(F)

FIG. 2.8. Typical valence bond structures for $PF_5$ based on $P^+$. Structures E and F are based on the square pyramidal and trigonal bipyramidal arrangements of ligands, respectively. The orbitals used by phosphorus for bonding are $sp_x$ hybrid and $p_y$, $p_z$ atomic orbitals in structure E, and $sp^2$ ($p_x$, $p_y$) hybrid and $p_z$ atomic orbital in structure F. Interatomic electron pairing is indicated by a line joining singly occupied orbitals, and orbitals containing lone-pairs of electrons are indicated by crossed hatching.

arrangement. The structure (G), based on the trigonal bipyramidal arrangement, with the three fluorine ligands in the same plane, and a lone pair in the $3p_z$ atomic orbital, is clearly not favoured because of the strong repulsion between the lone pair in the $sp^2$ hybrid and the $F^-$ ion to which it is directed (see Fig. 2.9). On the other hand, if the two lone pairs are placed in $sp$ hybrids,

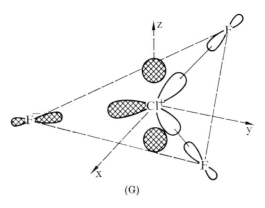

(G)

FIG. 2.9. A typical valence bond structure for $ClF_3$ based on $Cl^+$. The orbitals used by chlorine for bonding are the $sp^2$ $(p_x, p_y)$ hybrid and $p_z$ atomic orbitals. Interatomic electron pairing is indicated by a line joining singly occupied orbitals, and orbitals containing lone pairs of electrons are indicated by crossed hatching.

perpendicular to the molecular plane, and one fluorine atom is bonded to each of the $3p$ orbitals in the plane, then, after transferring one electron from chlorine to the remaining fluorine atom, there remains the problem of deciding where to place the $F^-$ ion. If it is placed *cis* to one of the two $p$-bonded fluorines, then it is possible to write several other equivalent structures which, after consideration of their mutual interaction, will lead to an overall stabilization of this basic $T$ structure (Fig. 2.10).

In this variant of the $T$ structure, the $3s$ atomic orbital is used entirely in forming the lone-pair orbitals, which project out perpendicular to the plane containing the fluorine atoms. An alternative T-shaped structure can be envisaged in which the four (valence) lone pair electrons are accommodated in two $sp^2$ hybrids (Fig. 2.11). Hence, just as in the case of $SF_4$, the molecular energy must be minimized to determine the optimum amount of $s$ character in the lone-pair hybrids. This calculation is expected to result in an energy minimized geometry in which one Cl—F bond is of a different length from the other two: a conclusion which is reached after noticing that the resonance between the structures (H) secures complete Cl—F bond equivalence, while resonance between the structures (I) imposes bond equivalence on only the two axial Cl—F bonds. In any event, the structure based on the use of valence $3s$

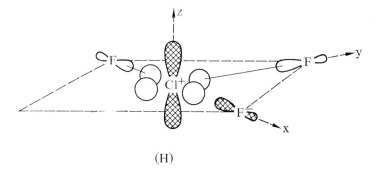

(H)

FIG. 2.10. A further possible valence bond structure for $ClF_3$ based on $Cl^+$. The orbitals used by chlorine for bonding are the $sp_z$ hybrid and $p_x$, $p_y$ atomic orbitals. Interatomic electron pairing is indicated by a line joining singly occupied orbitals, and orbitals containing lone-pairs of electrons are indicated by crossed hatching.

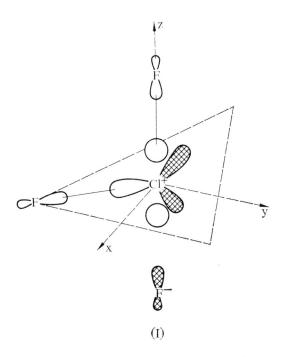

(I)

FIG. 2.11. Another typical valence bond structure for $ClF_3$ based on $Cl^+$. The orbitals used by chlorine for bonding are the $sp^2$ ($p_x$, $p_y$) hybrid and $p_z$ atomic orbitals. Interatomic electron pairing is indicated by a line joining singly occupied orbitals, and orbitals containing lone-pairs of electrons are indicated by crossed hatching.

and $3p$ atomic orbitals of chlorine will have a T-shaped form, with the F–Cl–F bond angles of 180° and 90°.

All the structures considered above for $ClF_3$ require the central chlorine atom to be singly ionized. But as the promotion energies for $Cl^+$, in any of the structures (H) or (I), are small (0·06 aJ for (H), and 0·70 aJ for (I)), and the energy required for charge separation is not too large, a favourable situation is created for regaining sufficient energy through bond formation to stabilize the molecule. Structures involving participation by $d$-orbitals can be added to improve the molecular wave function, by reducing the amount of charge transfer formally required. The calculation is a complicated one within the framework of valence bond theory but, in parallel with the results mentioned earlier, the available semi-empirical molecular orbital calculations indicate only a small contribution of structures involving $d$-orbitals.[15–16]

The main conclusion of this chapter is that it is quite possible to understand the geometries of second-row compounds without involving the participation of $d$-orbitals. The simple valence bond model, which is used here, shows how special properties of both second-row atom and ligands are required in order to achieve molecular stability, in contrast to the simple Pauling–Slater approach which considered the electron configuration of the central atom to be of prime importance. This simple model can be refined by extending the set of valence shell atomic orbitals to include $d$ functions: but, it is important to notice that by this procedure, the admixture of higher energy orbitals is determined variationally, in contrast to the simple Pauling–Slater hybrid scheme.

The simple valence bond model will be used again to discuss the electronic structure of rare gas compounds. But, before this is done, it is more suitable to discuss the electronic structure of transition metal ion complexes.

## REFERENCES

1. PAULING, L., *J. Amer. Chem. Soc.*, 1931, **53**, 1367, 3225.
2. SLATER, J. C., *Phys. Rev.*, 1931, **37**, 841.
3. McWEENY, R., *Symmetry—An Introduction to Group Theory and its Applications* (Topic 1, volume 3 of this series), Pergamon Press, Oxford (1963).
4. SLATER, J. C., *Quantum Theory of Atomic Structure*, volumes I and II, McGraw-Hill Book Co. Inc., New York (1960).
5. HINZE, J. and JAFFÉ, H. H., *J. Amer. Chem. Soc.*, 1962, **84**, 540.
6. CRAIG, D. P. and THIRUNAMACHANDRAN, T., *J. Chem. Phys.*, 1966, **45**, 3355.
7. COULSON, C. A. and GIANTURCO, F. A., *J. Chem. Soc.*, 1968, A, 1618.
8. CRAIG, D. P. and ZAULI, C., *J. Chem. Soc.*, 1962, **37**, 601.
9. MITCHELL, K. A. R., *J. Chem. Soc.*, 1968, A, 2676.
10. CRUICKSHANK, D. W. J., WEBSTER, B. C. and SPINNLER, M. A., *Intern. J. Quantum Chem.*, 1967, **1S**, 225.
11. LINDERBERG, J., *Intern. J. Quantum Chem.*, 1967, **1S**, 231.
12. RUNDLE, R. E., *Survey Prog. Chem.*, 1963, **1**, 81.
13. BROWN, R. D. and PEEL, J. B., *Austr. J. Chem.*, 1968, **21**, 2605.
14. HILLIER, I. H., *J. Chem. Soc.*, 1969, A, 878.
15. SANTRY, D. P. and SEGAL, G. A., *J. Chem. Phys.*, 1967, **47**, 158.

16. BROWN, R. D. and PEEL, J. B., *Austr. J. Chem.*, 1968, **21**, 2617.
17. WILLETT, R. D., *Theor. Chim. Acta*, 1964, **2**, 393.
18. ROBERTS, H. L., *Inorganic Sulphur Chemistry*, 1968, chapter 12, Elsevier, Amsterdam.
19. BERRY, R. S., TAMRES, M., BALLHAUSEN, C. J. and JOHANSEN, H., *Acta Chem. Scand.*, 1968, **22**, 231.
20. BERRY, R. S., *J. Chem. Phys.*, 1960, **32**, 933.
21. BERRY, R. S., *Rev. Mod. Phys.*, 1960, **32**, 447.
22. GILLESPIE, R. J., *J. Chem. Soc.*, 1963, 4672.

# BIBLIOGRAPHY

MITCHELL, K. A. R., The use of outer $d$ orbitals in bonding, *Chem. Rev.*, 1969, **69**, 157.

COULSON, C. A., $d$-electrons and molecular bonding, *Nature*, 1969, **221**, 1106.

CRUICKSHANK, D. W. J. and WEBSTER, B. C., Orbitals in sulphur and its compounds, *Inorganic Sulphur Chemistry*, 1968, chapter 2, Elsevier, Amsterdam.

MUSHER, J. I., The chemistry of hypervalent molecules, *Angew. Chem. Internat. Edit.*, 1969, **8**, 54.

KIMBALL, G. E., Directed valence, *J. Chem. Phys.*, 1940, **8**, 188.

PAULING, L., *Nature of the Chemical Bond*, 1942, Cornell University Press.

CHANDLER, G. S. and THIRUNAMACHANDRAN, T. (with an Appendix by CAMPBELL, J. A. and CHANDLER, G. S.): $d$ orbitals in the $sp^3d$, $s^1p^2d^2$, and $p^3d^2$ configurations of phosphorus, *J. Chem. Phys.*, 1968, **49**, 3640.

SCHONLAND, D., *Molecular Symmetry*, 1965, D. Van Nostrand Co. Ltd., London.

LUCKEN, E. A. C., Valence-shell expansion studied by radio-frequency spectroscopy, *Structure and Bonding*, 1969, **6**, 1.

JØRGENSEN, C. K., Valence-shell expansion studied by ultra-violet spectroscopy, *Structure and Bonding*, 1969, **6**, 94.

MARSMANN, H., VAN WAZER, J. R. and ROBERT, J.-B., $d$ orbitals in positive, neutral and negative phosphorus atoms, *J. Chem. Soc.*, 1970, A, 1566.

CRAIG, D. P. and MACLAGAN, R. G. A. R., $d$ orbitals in excited configurations of chlorine, *J. Chem. Soc.*, 1970, A, 1431.

# CHAPTER 3

# THE ELECTRONIC STRUCTURE OF TRANSITION METAL ION COMPLEXES

## 3.1. Introduction

The crystal field model for discussing the electronic structure of transition metal ion complexes was proposed by Bethe[1] in 1929. In this model, the ligands are usually simulated by suitably chosen point charges, each one of which acts as a source of Coulombic potential, and no allowance is made for explicit involvement of ligand atomic orbitals. Thus, the wave function for the complex corresponds, in effect, to a valence-bond ionic structure, in which the detailed electronic structure of the ligands is ignored. The theory was developed primarily for understanding the spectral properties of a transition metal ion, or rare earth ion, embedded in an ionic matrix; for example, $Cr^{+++}$ embedded in $Al_2O_3$ (ruby), where each $Cr^{+++}$ ion is surrounded by six nearest-neighbour $O^{--}$ ions at an average distance of about $0.19$ nm.

The apparent success of the ionic model led to its use in correlating the spectral properties of a wide range of transition metal and rare earth ion complexes. But the applications have tended to go beyond the bounds of the original theory, and it is not surprising to find that a detailed understanding of transition metal-ion complexes must necessarily take cognizance of the electronic structure of the ligands. Nevertheless, it is still very informative to develop the formalism of crystal field theory because, as will be seen later, the theory can be generalized to deal with situations in which ligand orbital participation is important. Although the ionic model has limitations, it does help in understanding the origin of the various terms contributing to the energy of the metal ion in the complex or crystal.

The crystal field Hamiltonian is given by the free ion Hamiltonian, $\hat{H}_0$, plus an additional potential energy term, $\hat{V}$, arising from the potential due to the array of point charges representing the ligands in a transition metal ion complex, or the ions of the host lattice. However, as this chapter is basically concerned with isolated transition metal ion complexes, $\hat{V}$ will possess the same symmetry as that of the array of ligands surrounding the central metal ion. Thus, for a transition metal ion with $n$ valence electrons, surrounded by

$N$ ligands,

$$\hat{H} = \sum_{i=1}^{n} (-\tfrac{1}{2}\nabla_i^2 + U_{\text{core}}(i)) + \sum_{i<j} \frac{1}{r_{ij}} + \sum_i \xi_i(\hat{\mathbf{l}}_i \cdot \hat{\mathbf{s}}_i) + \sum_{\alpha=1}^{N} \sum_{i=1}^{n} \hat{V}_\alpha(i)$$

$$= \hat{H}_1 + \hat{H}_2 + \hat{H}_3 + \hat{V}$$

$$= \hat{H}_0 + \hat{V}$$

where, for simplicity, the only relativistic term retained is the one correspond-ing to the effects of spin-orbit coupling for the central metal ion.

$$\sum_{\alpha=1}^{N} \hat{V}_\alpha(i) \qquad \text{and} \qquad U_{\text{core}}(i)$$

are the potential energies of electron $i$ in the fields provided by the ligands and the metal ion core, respectively. It is convenient to refer to the four terms in $\hat{H}$ as $\hat{H}_1$, $H_2$, $\hat{H}_3$, and $\hat{V}$, respectively.

The basic problem is to find a suitable set of wave functions for use in describing the (approximate) solutions of the Schrödinger equation with Hamiltonian, $\hat{H}$. As already explained in Chapter 1, the solution is best approached in a step-wise manner when the Hamiltonian consists of several terms: for example, $\hat{H}_1 + \hat{H}_2$ could be diagonalized with respect to a chosen set of expansion functions, and then the solutions used for the successive diagonalization of $\hat{H}_3$ and then $\hat{V}$. The most convenient choice for the initial set of expansion functions† depends on the relative importance of $\hat{V}$, $\hat{H}_2$ and $\hat{H}_3$. For the "weak field" situation, where the electron–electron interaction term, $\hat{H}_2$, is more important than $\hat{V}$, which is in turn more important than $\hat{H}_3$, it is advisable to construct the initial set of wave functions from those functions which diagonalize $\hat{H}_1 + \hat{H}_2$: the effects of $\hat{V}$ are then small, and the overall pattern of energy levels will not be disturbed too much. On the other hand, in the "strong field" situation, where the effects of $\hat{H}_3$ are less than the effects of $\hat{H}_2$ which are in turn less than the effects of $\hat{V}$, the convenient expansion functions are those which diagonalize $\hat{H}_1 + \hat{V}$. There are several other choices for the initial set of expansion functions, depending upon the relative importance of the three terms in $\hat{H}$ under consideration.

In the weak field case, where the effects of spin-orbit coupling are usually relatively unimportant, the zeroth order wave functions are given by the set of atomic term wave functions arising from the various valence electron configurations, appropriate to the atom of interest—these are the wave functions which diagonalize $\hat{H}_1 + \hat{H}_2$. The neglect of spin-orbit coupling therefore implies the applicability of Russell–Saunders coupling, in so far as $L$ and $S$ are considered to remain good quantum numbers at this stage in the calculation. As an example, consider the terms arising from the valence

---

† It may be more convenient to allow for nuclear screening effects as indicated in the footnote on p. 8.

electron configurations $3d^2$, $3d4s$ and $4s^2$ of $V^{+++}$:

$$3d^2 : {}^1S, \ {}^1G, \ {}^3P, \ {}^1D, \ {}^3F,$$

$$3d4s : {}^1D, \ {}^3D,$$

$$4s^2 : {}^1S.$$

It is often sufficient to consider only the terms arising from the configuration $3d^2$, on account of the large separation in energy between these terms and the terms arising from the configurations $3d4s$ and $4s^2$.

Since each term is $(2L + 1)(2S + 1)$-fold degenerate, the three configurations give rise to a total of sixty-six component wave functions. However, these wave functions themselves do not form a suitable set of zeroth order wave functions, unless the expansion is restricted to the wave functions arising from the dominant configuration, $3d^2$. The occurrence of both $^1D$ and $^1S$ terms in different configurations means that, with the use of orbital approximations to the component wave functions of these terms, the $66 \times 66$ secular determinant will not be diagonal with respect to $\hat{H}_1 + \hat{H}_2$. This result follows because the orbital approximations to the exact term wave functions are not eigenfunctions of $\hat{H}_1 + \hat{H}_2$, and hence some allowance must be made for the interaction between approximate term wave functions of the same symmetry: that is, the optimum $^1D$ and $^1S$ wave functions, for a given $M_L$, are obtained from the solution of the secular equations

$$C_{1X}(H_{11} - E) + C_{2X}H_{12} = 0,$$

$$C_{1X}H_{21} + C_{2X}(H_{22} - E) = 0,$$

where

$$H_{ij} = \int \Psi_i(^1X)(\hat{H}_1 + \hat{H}_2)\Psi_j(^1X) \, d\tau$$

with $X = D$ or $S$.

If there is a large amount of configurational mixing, then it may be difficult to ascribe a unique electron configuration to the term in question. An interesting example of this situation has been observed by Zare[2] in calculations on $Al^+$, where the $^1D$ term, formally arising from the valence electron configuration $3s3d$, cannot be associated with a unique electron configuration:

$$\Psi(^1D) = 0{\cdot}66(3s3d) - 0{\cdot}52(3s4d) + 0{\cdot}49(3p3p) - 0{\cdot}16(3s5d) + \ldots .$$

Configurational mixing is usually ignored in calculations involving transition metal ions, and it is hoped that the term wave functions are adequately represented by antisymmetrized products of orbitals associated with the dominant electron configurations. The inclusion of $\hat{V}$, the crystal field Hamiltonian, will, in general, cause a coupling between the free atom term wave functions with the same spin multiplicity. If the effect of the crystal field is weak, the off-diagonal coupling terms will be small and, in this situation, the problem may be safely reduced to one involving the lower energy terms arising from the $3d^2$ valence electron configuration. The influence of terms, arising from higher energy configurations, can be treated by means of perturbation theory, since their effect is proportional to the square of the appropriate matrix element of $\hat{V}$ divided by the difference in term energies; and this ratio should be small for weak fields.

The coupling between the free atom states arises because, in the field of the ligands, the orbital angular momentum does not remain a constant of the motion (the solutions of Schrödinger's equation for the complex are not eigenfunctions of $\hat{L}^2$ because of the loss of central field symmetry). Thus, in the presence of the ligand field, $L$ is no longer a good quantum number for characterizing the electronic state of the central metal ion: in fact, the set of $(2L + 1)$ component wave functions, associated with each term, may no longer remain degenerate. To be more specific: the crystal field induces a coupling between component wave functions possessing the same spatial and spin symmetries under the group of symmetry operations characterizing the complex ion. If the effect of $\hat{H}_3$ is neglected, $S$ remains a good quantum number, then, for example, $\hat{V}$ couples the component wave functions associated with the $^3F$ and $^3P$ terms of $3d^2$ in environments of octahedral and tetrahedral symmetry. The actual magnitudes of the term splittings can be found by working out the various coupling terms, and then diagonalizing $\hat{H}$.

Unfortunately, this approach has several disadvantages. First, as the coupling terms increase in magnitude with increasing crystal field strength, it may become necessary to include terms arising from higher energy configurations—that is, the effects of $\hat{V}$ are becoming comparable with those of $\hat{H}_2$. Secondly, and more important, the basic one-electron orbitals, used for constructing approximate term wave functions, are eigenfunctions of a problem with a spherically symmetrical potential. Clearly, it would be more satisfactory if the symmetry possessed by the crystal field was taken into account from the start; the one-electron orbitals could then be chosen to display the symmetry properties of $G$, rather than those of the full rotation group. The advantage gained from this approach is that both crystal field and molecular orbital theories (discussed later) can be developed within the same framework. This arises because in the molecular orbital treatment of a closed-shell molecule—open-shell molecules require further consideration—the molecular orbitals can be classified according to the symmetry group of the effective one-electron Hamiltonian, $\hat{h}$; normally the same group as $G$. Thus, the first problem in this "strong field" approach is to examine the transformation properties of the various atomic orbitals under finite groups of symmetry operations: in particular, the octahedral and tetrahedral symmetry groups, in view of the widespread number of complexes displaying these point group symmetries.

Now it has already been shown in Chapter 2 that $s$, $p$ and $d$ atomic orbitals transform according to the irreducible representations $a_1$, $t_2$ and $e + t_2$, respectively, of $T_d$, or the irreducible representations $a_{1g}$, $t_{1u}$ and $e_g + t_{2g}$, respectively, of $O_h$. It only remains, therefore, to discuss the transformation properties of $f$ atomic orbitals.

The matrix representatives of the group generators are now of order seven and, on aplying the same techniques which were used in dealing with the set of five $d$ orbitals, the reducible representation is found to contain the irreducible representations $a_2$, $t_1$, $t_2$ under $T_d$, and $a_{2u}$, $t_{1u}$, $t_{2u}$ under $O_h$. An interesting situation now arises for, in tetrahedral fields, the sets of $p_x$, $p_y$, $p_z$; $d_{xy}$, $d_{xz}$, $d_{yz}$; and $f_{x^3}$, $f_{y^3}$, $f_{z^3}$ atomic orbitals all span the irreducible representation $t_2$. This makes it impossible, for example, to separate the relative contributions of central metal $3d$, $4p$ and $4f$ atomic orbitals to the molecular bonding in a tetrahedral complex. However, for metal atoms in the first transition series, where the $4f$ atomic orbitals are energetically inaccessible, it is common practice to describe the metal atom contribution to the bonding in terms of $3d$ and $4p$ atomic orbitals only.

A similar situation obtains in octahedral complexes, but now it is the sets of $p_x$, $p_y$, $p_z$ and $f_{x^3}$, $f_{y^3}$, $f_{z^3}$ atomic orbitals which span the same irreducible representation, $t_{1u}$. Once again, in the case of an atom in the first transition series, the contribution to the bonding from $4f$ atomic orbitals is usually ignored. The extra symmetry possessed by the octahedral ligand grouping therefore allows the central metal $p$ and $d$ contributions to the bonding to be considered separately. The lack of an analogous separation in tetrahedral complexes makes calculations on these compounds more difficult: a conclusion which is amply supported by the existence, in the literature, of many conflicting calculations—particularly on the $MnO_4^-$ ion.

Although the mixing between central atom $p$ and $f$ atomic orbitals can usually be neglected in first-row transition metal ion complexes, special care is needed when discussing compounds containing a central rare earth ion. For example, the $Tm^{++}$ ion in $CaF_2$ is in a site of local cubic symmetry and, with the $4f$ shell already occupied, it is inadvisable to treat the crystal field splitting of the set of $4f$ atomic orbitals without including participation of $5p$ atomic orbitals.

The group theoretical discussion given above shows how the degeneracy of the free metal ion $d$ and $f$ atomic orbitals is partially removed when the ion is placed in an environment of reduced symmetry. But, it is important to realize that the group theory can give no quantitative information on the details of the orbital splitting pattern; this information can only be obtained from the results of model calculations.

As discussed in the beginning of this chapter, the model provided by the crystal field theory was used in the early theoretical development of transition metal ion chemistry. However, more recent work tends to favour a different model based on the use of some form of molecular orbital or valence bond theory. It is therefore useful to discuss, and compare, these models in some detail, in order that their respective limitations can be readily ascertained.

## 3.2. The crystal field model: $d^1$ complexes

In the strong field approach to crystal field theory, the zeroth-order Hamiltonian†

$$\hat{H}_1 + \hat{V} \tag{3.1}$$

is invariant only to the symmetry operations of the molecular point group, $G$, in contrast to the zeroth-order Hamiltonian used in the weak field approach. Thus, the eigenfunctions of $\hat{H}_1 + \hat{V}$ are labelled by the irreducible representations of $G$, rather than by $L$ and $S$.

Since (3.1) is a sum of one-electron operators, the simple (antisymmetrized) products of the appropriate one-electron orbitals will form a suitable set of zeroth-order expansion functions. For example, the allocation of $n$-electrons to the $t_{2g}$ and $e_g$ orbitals, in an octahedral environment, leads to electron configurations of the kind $t_{2g}^m e_g^{n-m}$ ($m = 0, 1, \ldots, n$) where, for a given value of $m$, several product wave functions can usually be constructed on account of the degeneracy associated with each of the basic orbitals. The complete set of wave functions, arising from all possible allocations of the electrons, diagonalizes (3.1); furthermore, all component wave functions associated with the same value of $m$ are degenerate.

The inclusion of $\hat{H}_2$ in the Hamiltonian removes the degeneracy of the component wave functions, associated with the same value of $m$. The diagonalization of $\hat{H}_2$ then leads to a set of molecular term energies, in exactly the same way as atomic term energies are derived from atomic electron configurations when the interelectronic interactions are included in the atomic Hamiltonian. It is perhaps useful, at this juncture, to consider the various molecular electronic states which can arise from $d^1$ and $d^2$ electron configurations of transition metal ions in the presence of an octahedral field: this should show the basic principles of the strong field approach more clearly.

A $d^1$ metal ion electron configuration arises in complexes of Ti$^{+++}$, like [Ti(H$_2$O)$_6$]$^{+++}$ or [TiF$_6$]$^{---}$. The two strong field configurations $t_{2g}^1$ and $e_g^1$ give rise to a total of ten one-electron component wave functions, of which five are associated with $M_S = +\frac{1}{2}$ and five with $M_S = -\frac{1}{2}$. Thus, for a given value of $M_S$, the five component wave functions may be grouped into the three component wave functions of the molecular $^2T_{2g}$ term, and the two component wave functions of the $^2E_g$ term—assuming that the effects of spin-orbit

† It may be more convenient to treat $\hat{H}_1 + \hat{V} + \hat{V}_{\text{eff}}$ as the zeroth-order Hamiltonian, and modify $\hat{H}_2$ accordingly (see the footnote to p. 8).

coupling are neglected. The energy matrix is diagonal with respect to these component wave functions:

$$H_{0a0a} \equiv E(^2T_{2g}) = \int \Psi_0[^2T_{2g}(t_{2g}); a, M_S = \tfrac{1}{2}][\hat{H}_1 + \hat{V}] \times$$
$$\times \Psi_0[^2T_{2g}(t_{2g}); a, M_S = \tfrac{1}{2}] \, d\tau \quad (a = 1, 2, 3),$$

$$H_{1b1b} \equiv E(^2E_g) = \int \Psi_1[^2E_g(e_g); b, M_S = \tfrac{1}{2}][\hat{H}_1 + \hat{V}] \times$$
$$\times \Psi_1[^2E_g(e_g); b, M_S = \tfrac{1}{2}] \, d\tau \quad (b = 1, 2),$$

$$H_{0a1b} = 0; \qquad H_{0a0a'} = 0; \qquad H_{1b1b'} = 0, \qquad (3.2)$$

where, for convenience, a slightly different notation has been introduced for the component wave functions of a given molecular term. The term symmetry, parent electron configuration, a label designating the component wave function of the term, and the value of $M_S$, respectively, are all given as arguments rather than $\mathbf{r}$, $\mathbf{R}$, $\mathbf{s}$ which are not given explicitly.

In the crystal field model, the three components of the $^2T_{2g}$ term, with $M_S = \tfrac{1}{2}$, are given by the three (spin) orbitals $d_{xz}$, $d_{yz}$, $d_{xy}$ of the central metal ion. However, for future discussions, it is convenient to use the notation introduced by Griffith,[3] whereby the component wave functions forming a basis for the irreducible representation $t_{2g}$ (or $T_{2g}$) are designated by $\eta$, $\xi$, and $\zeta$, respectively. Similarly, the two functions forming a basis for the irreducible representation $e_g$ (or $E_g$) are labelled by $\theta$ and $\epsilon$: in the present situation, these label the strong field orbitals previously designated by $d_{z^2}$ and $d_{x^2-y^2}$, respectively.

An estimate of the magnitude of the separation in energy between the $^2T_{2g}$ and $^2E_g$ molecular states requires explicit evaluation of the matrix elements appearing in (3.2). This is best accomplished by first considering the crystal field contribution to the Hamiltonian:

$$\hat{V} = \sum_\alpha \sum_m \hat{V}_\alpha(m) = -Q \sum_\alpha \sum_m \frac{1}{r_{\alpha m}}$$

$Q$ is the net ligand charge (in units of $e$), and $r_{\alpha m}$ is the distance of electron $m$ from ligand $\alpha$ (only charged monoatomic ligands are considered here; the form of $\hat{V}$ must be generalized when dealing with neutral or charged polyatomic ligands).

$\hat{V}$ therefore gives rise to a number of two-centre integrals, each of which is usually reduced to a sum of one-centre integrals by expressing $r_{\alpha m}$ in terms of distances and angles relating to the coordinate system centred on the metal ion:

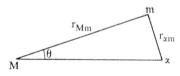

Thus, on application of the cosine law,

$$r_{\alpha m} = \{r_{Mm}^2 + r_{M\alpha}^2 - 2r_{Mm}r_{M\alpha}\cos\theta\}^{1/2}$$

where, as shown in the above figure, $r_{Mm}$ and $r_{M\alpha}$ are the distances between electron $m$ and the metal ion nucleus, and between nucleus $\alpha$ and the metal ion, respectively; $\theta$ is the angle between $r_{M\alpha}$ and $r_{Mm}$, and the line joining the nuclei $M$ and $\alpha$ is taken as the axis of quantization ($z$-axis). Hence,

$$1/r_{\alpha m} = \frac{1}{r_{Mm}}\left\{1 + \left(\frac{r_{M\alpha}}{r_{Mm}}\right)^2 - 2\left(\frac{r_{M\alpha}}{r_{Mm}}\right)\cos\theta\right\}^{-1/2}, \quad r_{Mm} > r_{M\alpha}$$

$$= \frac{1}{r_{M\alpha}}\left\{1 + \left(\frac{r_{Mm}}{r_{M\alpha}}\right)^2 - 2\left(\frac{r_{Mm}}{r_{M\alpha}}\right)\cos\theta\right\}^{-1/2}, \quad r_{Mm} < r_{M\alpha},$$

and, on expanding the expression in curly brackets in terms of Legendre polynomials (see Margenau and Murphy,[4] p. 100), the expression for $1/r_{\alpha m}$ becomes

$$1/r_{\alpha m} = \frac{1}{r_>}\sum_{n=0}^{\infty}\left(\frac{r_<}{r_>}\right)^n P_n(\cos\theta),$$

where $r_<$, $r_>$ represents the smaller or larger of $r_{Mm}$, $r_{M\alpha}$, respectively.

The Legendre polynomial $P_n(\cos\theta)$, with argument determined by the coordinate system on $M$, in which the $z$-axis is directed towards $\alpha$, can always be transformed to allow for an arbitrary inclination of the $z$-axis to the chosen $M$—$\alpha$ bond:

$$P_n(\cos\theta) = P_n(\cos\theta_{Mm})P_n(\cos\theta_{M\alpha}) + 2\sum_{t=1}^{n}\frac{(n-t)!}{(n+t)!}P_n^t(\cos\theta_{Mm}) \times$$

$$\times P_n^t(\cos\theta_{M\alpha})\cos t(\phi_m - \phi_\alpha),$$

where $\theta_{Mm}$, $\phi_m$; $\theta_{M\alpha}$, $\phi_\alpha$ are the angles defining the orientation of $r_{Mm}$ and $r_{M\alpha}$, respectively, to the new polar axis.

The crystal field Hamiltonian is now given by the following sum of one-centre terms:

$$\hat{V} = -\frac{Q}{r_>}\sum_{m}^{N}\sum_{\alpha=1}^{\infty}\sum_{n=0}\left(\frac{r_<}{r_>}\right)^n\left[P_n(\cos\theta_{Mm})P_n(\cos\theta_{M\alpha}) + 2\sum_{t=1}^{n}\frac{(n-t)!}{(n+t)!}\times\right.$$

$$\left.\times P_n^t(\cos\theta_{Mm})P_n^t(\cos\theta_{M\alpha})\cos t(\phi_m - \phi_\alpha)\right]$$

which, in the case of an octahedral field, becomes

$$\hat{V} = \sum_{m}\left[-\frac{6Q}{r_>} - Q\frac{r_<^4}{r_>^5}\left\{\frac{7}{2}P_4^0(\cos\theta_{Mm}) + \frac{1}{48}P_4^4(\cos\theta_{Mm})\cos4\phi_m\right\}\right].$$

Terms arising from values of $n$ greater than 4 are not given explicitly, as they yield a zero contribution to matrix elements involving $d$-orbitals.

Hence, after integrating over the angular coordinates, the diagonal matrix element of $\hat{V}$, for the $\theta$-component of the $^2E_g$ term, is given by

$$\int_0^{r_{M\alpha}} g_{d\theta}(r_{Mm})g_{d\theta}(r_{Mm})\left[-\frac{6Q}{r_{M\alpha}} - Q\frac{r_{Mm}^4}{r_{M\alpha}^5}\right] r_{Mm}^2\, \mathrm{d}r_{Mm}$$

$$+ \int_{r_{M\alpha}}^\infty g_{d\theta}(r_{Mm})g_{d\theta}(r_{Mm})\left[-\frac{6Q}{r_{Mm}} - Q\frac{r_{M\alpha}^4}{r_{Mm}^5}\right] r_{Mm}^2\, \mathrm{d}r_{Mm}$$

$$= \int_0^\infty [g_{d\theta}(r_{Mm})]^2 \left[-\frac{6Q}{r_{M\alpha}} - Q\frac{r_{Mm}^4}{r_{M\alpha}^5}\right] r_{Mm}^2\, \mathrm{d}r_{Mm}$$

$$- \int_{r_{M\alpha}}^\infty [g_{d\theta}(r_{Mm})]^2 \left[-\frac{6Q}{r_{M\alpha}} - \frac{Qr_{Mm}^4}{r_{M\alpha}^5} + \frac{6Q}{r_{Mm}} + Q\frac{r_{M\alpha}^4}{r_{Mm}^5}\right] \times$$

$$\times r_{Mm}^2\, \mathrm{d}r_{Mm}, \quad (3.3)$$

where $g_{d\theta}(r_{Mm})$ is the radial part of the $3d_{z^2}$ atomic orbital.

The second integral in (3.3) has a very small value, so long as $g_{d\theta}(r_{Mm})$ has negligible amplitude in the region $r_{M\alpha} < r_{Mm} < \infty$: a condition which is usually satisfied for multiply charged transition metal ions, on account of the compact form of the $3d$ orbitals. In fact, if the $d$-orbital is represented by a single Slater orbital, with exponent $k$, then the magnitude of the second integral is given approximately by $e^{-2kr_{M\alpha}}$ times the value of the first integral: that is, for values of $k$ appropriate to a multiply charged transition metal ion (about 6), the contribution of the second integral is negligible. Hence, the required matrix element can be written in the form

$$-\frac{6Q}{r_{M\alpha}} - \frac{Q}{r_{M\alpha}^5} \int_0^\infty [g_{d\theta}(r_{Mm})]^2\, r_{Mm}^4 . r_{Mm}^2\, \mathrm{d}r_{Mm}$$

$$= -\frac{6Q}{r_{M\alpha}} - \frac{Q}{r_{M\alpha}^5} \langle r_{Mm}^4 \rangle_{e_g} = C - \frac{Q}{r_{M\alpha}^5} \langle r_{Mm}^4 \rangle_{e_g}$$

where $C$ is a constant, independent of orbital size, and $\langle r_{Mm}^4 \rangle_{e_g}$ is the average value of $r_{Mm}^4$ for the $3d_{z^2}$ radial function.

The complete matrix element is therefore given by

$$\int \Psi[^2E_g(e_g); \theta, M_S = \tfrac{1}{2}][\hat{H}_1 + \hat{V}]\Psi[^2E_g(e_g); \theta, M_S = \tfrac{1}{2}]\, \mathrm{d}\tau_1$$

$$= \gamma + C - \frac{Q}{r_{M\alpha}^5} \langle r_{Mm}^4 \rangle_{e_g},$$

where $\gamma$ represents the energy of the $d_{z^2}$ electron in the field of the nucleus and core electrons of the metal ion.

The energy of each component of the $^2T_{2g}$ state follows in a similar way: for example,

$$\int \Psi[^2T_{2g}(t_{2g}); \xi, M_S = \tfrac{1}{2}][\hat{H}_1 + \hat{V}]\Psi[^2T_{2g}(t_{2g}); \xi, M_S = \tfrac{1}{2}]\, \mathrm{d}\tau_1$$

$$= \epsilon + C + \frac{2}{3}\frac{Q}{r_{M\alpha}^5} \langle r_{Mm}^4 \rangle_{t_{2g}}$$

where

$$\epsilon = \int \Psi[^2T_{2g}(t_{2g}); \xi, M_S = \tfrac{1}{2}]\hat{H}_1\Psi[^2T_{2g}(t_{2g}); \xi, M_S = \tfrac{1}{2}]\,d\tau_1.$$

Now in an ion like Ti$^{+++}$, with compact $3d$ orbitals, the variation of $3d$ orbital size with strength of crystal field is only going to be significant for strong fields. But, in this situation, the whole approach breaks down, because it will then be totally inadequate to represent the ligands by point charges. Under normal circumstances, therefore, it seems reasonable to associate the $t_{2g}$ and $e_g$ metal ion orbitals with the same radial function, $g_d(r_{Mm})$. The number of parameters is then reduced to two, since

$$\gamma = \epsilon, \text{ and } \langle r^4_{Mm}\rangle_{e_g} = \langle r^4_{Mm}\rangle_{t_{2g}} = \langle r^4\rangle$$

say, and the separation in energy between the $^2T_{2g}$ and $^2E_g$ molecular states is given by

$$E(^2E_g) - E(^2T_{2g}) = -\tfrac{5}{3}Q\langle r^4\rangle/r^5_{M\alpha} = \Delta.$$

For traditional reasons, this energy separation is often referred to as $10Dq$, where

$$D = -\frac{35Q}{4r^5_{M\alpha}} \quad \text{and} \quad q = \frac{2}{105}\langle r^4\rangle.$$

However, it is often difficult to select unambiguous values for the crystal field parameters $D$ and $q$ (or $Q$ and $\langle r^4\rangle$), and so $\Delta$ is usually regarded as an empirical parameter which can be chosen by appeal to experiment.

The results of the crystal field calculation are shown schematically in Fig. 3.1. The figure is drawn in such a way to show the origin of the various terms contributing to the energy of the metal ion valence electron in the complex, when the ligands are simulated by point negative charges. First of all, the energy of the electron arising from the spherically symmetrical part of the crystal potential results in a uniform shift of all $d$-orbital energies by an amount $C$. The levels are then split by the only non-totally symmetrical component of $\hat{V}$ which is of physical significance. If the splitting is large, the optimum radial forms for the $t_{2g}$ and $e_g$ orbitals will be significantly different

FIG. 3.1. A schematic representation of the crystal field splitting of the free ion $3d$ atomic orbitals, which is induced by an octahedral array of point negative charges.

(the energy of the $t_{2g}^1$ configuration will be minimized for a different value of $k$ than that of the $e_g^1$ configuration, assuming that Slater orbitals are used), and this will result in an extra contribution to $\Delta$, arising from the difference in core-valence electron interaction energies, $\gamma-\epsilon$. This point is made only as a reminder, to show that $\Delta$ is best treated as an empirical parameter, as its magnitude cannot be ascribed solely to one particular source.

The magnitude of $\Delta$ can be determined from the experimentally observed energy of the electronic transition $^2E_g \leftarrow {}^2T_{2g}$. This transition is electric-dipole forbidden, when the nuclear framework possesses the regular octahedral geometry, as there is then no change in parity on excitation. In fact, the transition appears with very low intensity as a result of vibrational motions within the molecule which do not preserve the centre of symmetry. Under these circumstances, the selection rule, based on a change in parity, is not valid, and transitions can occur when the ligands are displaced from their equilibrium positions. This vibrational–electronic (vibronic) interaction can be allowed for by including an appropriate potential energy term, as a perturbation, in the molecular electronic Hamiltonian. The perturbation vanishes at the equilibrium nuclear configuration but, for other nuclear configurations, it causes a coupling between molecular electronic states. The previously inaccessible upper electronic state therefore becomes contaminated with small amounts of higher energy electronic states—for example, the $^2T_{1u}$ state which arises through the excitation of the metal electron into the $4p(t_{1u})$ shell—which give rise to small, but non-zero, contributions to the transition moment integral and, hence, to the intenisty of the absorption band.

The same mechanism of vibronic interaction is responsible for the appearance of $d$–$d$ transitions in other octahedral transition metal ion complexes, in so far as they involve the excitation of a single electron. It is therefore of some interest to examine the origin of the intensities of these formally forbidden transitions in more detail. But, before doing this (see Section 3.3), the discussion of the $d^1$ absorption spectra is first completed.

In the case of the hydrated $Ti^{+++}$ ion, $[Ti(H_2O)_6]^{+++}$, the maximum in the absorption band intensity is found to occur at 0·403 aJ. A further examination of the band envelope reveals a shoulder, to longer wavelength, which is indicative of the presence of another electronic transition. This is often taken as evidence for the manifestation of the Jahn–Teller effect; a conclusion which is reasonable in view of the orbital degeneracies of both upper and lower electronic states, but difficult to substantiate quantitatively. The $^2E_g$ electronic state is expected to be more susceptible to the coupling between electronic and vibrational motions, as the energy of the electron in an $e_g$ orbital, which is directed towards the ligands, depends more sensitively upon the ligand configuration than the energy of the electron in a $t_{2g}$ orbital. In fact, calculations made by Van Vleck[5] show quite clearly that the Jahn–Teller effect is expected to be of greater importance in the excited $^2E_g$ electronic state. The appearance of two overlapping absorption bands in the electronic spectrum of $[Ti(H_2O)_6]^{+++}$ is therefore associated with transitions to each one of the Jahn–Teller components of the upper $^2E_g$ electronic state. But, because of the difficulty in reliably assessing the presence of a Jahn–Teller effect—to observe it satisfactorily requires the distortion energy, induced by the vibronic coupling, to be greater than the zero-point energy for vibrational motion—the following discussion will be simplified by excluding the possibility of Jahn–Teller activity (and spin-orbit coupling). It must be remembered, though, that systems with degenerate electronic states may require a separate, and more detailed, discussion which includes the effects of vibronic and spin-orbit coupling.

The $[TiF_6]^{---}$ species has also been examined, and is found to give a low intensity absorption band with a maximum at about 0·34 aJ. But since these observations refer to $[TiF_6]^{---}$ units in crystalline solids, the $d$-orbital splitting will be influenced, to some extent, by the disposition of the other ions in the lattice. This result is, therefore, really outside the scope of the present chapter but, so long as the potential arising from the surrounding ions is uniform in the vicinity of the $[TiF_6]^{---}$ unit, the above value of $\Delta$ will give a fairly reliable measure of the crystal field splitting.

In the case of $d^1$ (or $d^9$) tetrahedral complexes, the $d$ orbital splitting pattern is inverted, yielding a $^2E$ (or $^2T_2$) ground electronic state. The absence of an inversion centre now makes the $^2T_2 \leftarrow {}^2E$ transition allowed. However, the increase in intensity of the absorption band can be satisfactorily accounted for only by allowing for the mixing of the $4p$ and $3d(t_2)$ metal ion orbitals, which occurs in a tetrahedral environment: if this mixing is ignored in the crystal field approach, then the intensity of the absorption band can result only from vibronic interactions, as in the case of the octahedral molecule. The calculations are further complicated by the presence of Jahn–Teller distortions in the ground state of the complex.

This completes the discussion of $d^1$ complexes and, as already mentioned above, it is now necessary to digress and examine, in more detail, the origin of the intensity associated with the formally forbidden $d$–$d$ electronic absorption bands. It has already been stated that the mechanism of vibronic interaction is responsible for the observed intensity in octahedral complexes, and it now remains to quantify the previous discussion.

## 3.3. The treatment of vibronic interactions

The ground and excited molecular wave functions of interest are represented by the product wave functions

$$\Phi_0 = \Psi_0(\mathbf{r}, \mathbf{R}, \mathbf{s})\Lambda_0^s(\mathbf{R}); \quad \Phi_1 = \Psi_1(\mathbf{r}, \mathbf{R}, \mathbf{s})\Lambda_1^t(\mathbf{R})$$

where, for example, $\Psi_0$, $\Psi_1$ represent the $^2T_{2g}$ and $^2E_g$ electronic states for the complex ion, and $\Lambda_i^k(\mathbf{R})$ represents the wave function for nuclear motion associated with the electronic state $\Psi_i$. In a more detailed treatment, the effects of Jahn–Teller coupling must be considered when either of the states $\Psi_0$ or $\Psi_1$ is orbitally degenerate. But these effects are neglected, for simplicity, in order to show the origin of the observed intensity in a "$g{\rightarrow}g$" transition.

The intensity of the electronic transition between $\Phi_0$ and $\Phi_1$ is governed by the magnitude of the transition moment vector, $\mathbf{M}$:

$$\mathbf{M} = \iiint \Phi_0 \left\{ -\sum_i \mathbf{r}_i + \sum_\alpha Z_\alpha \mathbf{R}_\alpha \right\} \Phi_1 \, d\mathbf{r} \, d\mathbf{R} \, d\mathbf{s}$$

$$= \mathbf{e}_1 M_x + \mathbf{e}_2 M_y + \mathbf{e}_3 M_z$$

where, in general, not all components of $\mathbf{M}$ are non-zero. Thus, on substituting for $\Phi_0$ and $\Phi_1$

$$\mathbf{M} = \iiint \Psi_0(\mathbf{r}, \mathbf{R}, \mathbf{s}) \Lambda_0^s(\mathbf{R}) \left\{ -\sum_i \mathbf{r}_i + \sum_\alpha Z_\alpha \mathbf{R}\alpha \right\} \Psi_1(\mathbf{r}, \mathbf{R}, \mathbf{s}) \Lambda_1^t(\mathbf{R}) \times$$
$$\times \, d\mathbf{r} \, d\mathbf{R} \, d\mathbf{s}$$

$$= \int \Lambda_0^s(\mathbf{R}) M_{01}(\mathbf{R}) \Lambda_1^t(\mathbf{R}) \, d\mathbf{R} \tag{3.4}$$

where

$$M_{01}(\mathbf{R}) = -\iint \Psi_0(\mathbf{r}, \mathbf{R}, \mathbf{S}) \left( \sum_i \mathbf{r}_i \right) \Psi_1(\mathbf{r}, \mathbf{R}, \mathbf{s}) \, d\tau$$

The term involving
$$\sum_\alpha Z_\alpha \mathbf{R}_\alpha$$

disappears, after integrating over the electronic coordinates, on account of the orthogonality of the two electronic wave functions.

In the present instance, where both $\Psi_0$ and $\Psi_1$ are assumed to have the same parity, $M_{01}$ vanishes at the equilibrium nuclear geometry for reasons of symmetry (the integrand changes sign under inversion). However, $M_{01}$ will have finite, but small, values for other nuclear configurations, and it remains to calculate the net contribution of these terms to the transition moment integral. The calculation is best made by expanding the electronic wave functions, for small departures from the equilibrium geometry, in terms of the set of electronic wave functions at the equilibrium nuclear geometry:

$$\Psi_k = \sum_j c_{jk}(\mathbf{R}) \Psi_j(\mathbf{r}, \mathbf{R}_0, \mathbf{s})$$

where the $c_{jk}$ are independent of the electronic coordinates; that is, the electronic wave function for the arbitrary nuclear configuration, $\mathbf{R}$, is obtained from the superposition of the electronic wave functions for the equilibrium nuclear geometry, $\mathbf{R}_0$. Although this procedure is conceptually useful, it is worth noting Liehr's[6] comments in respect of the possible slow convergence of the expansion in actual calculations.

Substitution for $\Psi'_0$ and $\Psi'_1$ in the expression for $M_{01}$ then yields

$$M_{01} = \sum_{j,\,k} c_{j0}(\mathbf{R})c_{k1}(\mathbf{R}) \int \Psi_j^*(\mathbf{r},\,\mathbf{R}_0,\,\mathbf{s}) \left( \sum_i \mathbf{r}_i \right) \Psi_k(\mathbf{r},\,\mathbf{R}_0,\,\mathbf{s})\, \mathrm{d}\tau$$
$$= \sum_{j,\,k} c_{j0}(\mathbf{R})c_{k1}(\mathbf{R})m_{jk}.$$

The $m_{jk}$ do not all vanish, as there will be terms where $j$, $k$ label states of opposite parity; this will result in $M_{01}$ having a finite value, which must then be averaged over the nuclear motions to yield $\mathbf{M}$. These particular pairs of electronic states, with opposite parity, are expected to be reasonably well separated in energy; and this enables perturbation theory to be used for determining the coefficients $c_{j0}$ and $c_{k1}$.

The perturbation is defined by the difference in molecular electronic Hamiltonians for the nuclear configurations $\mathbf{R}$ and $\mathbf{R}_0$. Thus, on taking the first two terms of the Taylor series expansion of $\hat{H}(\mathbf{R})$ about $\mathbf{R}_0$

$$\hat{H}(\mathbf{R}) = \hat{H}(\mathbf{R}_0) + \sum_i (\mathbf{R}_0^i - \mathbf{R}^i) \cdot (\nabla_{\mathbf{R}^i}\hat{H})_{\mathbf{R}=\mathbf{R}_0}$$

the perturbation is given by

$$\sum_i (\mathbf{R}_0^i - \mathbf{R}^i) \cdot (\nabla_{\mathbf{R}^i}\hat{H})_{\mathbf{R}=\mathbf{R}_0}.$$

The vector $\mathbf{R}^i - \mathbf{R}_0^i$ represents the displacement of nucleus $i$ from its equilibrium position. If this displacement is written as $\mathbf{q}^i$, then the electronic Hamiltonian for the distorted configuration becomes

$$\hat{H} = \hat{H}(0) + \sum_i \mathbf{q}^i \cdot (\nabla_{\mathbf{q}^i}\hat{H})_{\mathbf{q}=0} \tag{3.5}$$

where $\hat{H}(0) \equiv \hat{H}(\mathbf{R}_0)$ is the Hamiltonian for the equilibrium nuclear geometry (all the $\mathbf{q}^i = 0$). The terms $(\nabla_{\mathbf{q}^i}\hat{H})_{\mathbf{q}=0}$ are independent of $\mathbf{q}^i$, and depend only upon the electron coordinates and the nuclear coordinates appropriate to the configuration $\mathbf{R}_0$.

Now, in the octahedral complex $XY_6$, the seven nuclear displacement vectors, $\mathbf{q}^i$, giving rise to twenty-one Cartesian coordinates relative to local axes centred on each atom in the molecule, can be replaced by twenty-one symmetry adapted combinations of displacement coordinates, by the standard group theoretical methods already used for finding symmetry adapted combinations of atomic orbitals; that is, the operator

$$\sum_{\hat{R}} \chi_\Gamma^*(\hat{R})\hat{R}$$

$$\left( \text{or the more powerful operator } \sum_{\hat{R}} [D_\Gamma(R)]_{ji}^* \hat{R} \right)$$

is applied to an arbitrary displacement vector instead of an arbitrary atomic orbital (see McWeeny,[7] chapter 6).

Three different combinations of symmetry coordinates are found to represent pure rotations of the complex about each of the three reference axes on the central metal ion. And after multiplying each displacement coordinate by the square root of the appropriate atomic mass, these combinations of symmetry coordinates become the normal coordinates for rotational motion, $Q_l$ ($l = 1, 2, 3$). Similarly, it is a relatively easy matter to discover the three combinations of symmetry orbitals corresponding to pure translations of the complex along each of the three reference axes. The fifteen vibrational normal coordinates, $Q_l$ ($l = 7, \ldots, 21$), describing the relative motion of the nuclei in which the centre of mass remains fixed, can each be expressed as a linear combination of appropriate symmetry coordinates—the precise values of the coefficients being determined once the potential energy function for nuclear motion is specified.[8]

Thus, any arbitrary displacement of the nuclei, represented by the set of $\mathbf{q}^i$, can be expressed in terms of the various normal coordinates of nuclear motion, corresponding to translation, rotation and vibration of the complex. The main advantage of this scheme is that the normal coordinates can be classified according to the irreducible representations of the point group associated with the equilibrium molecular geometry; and this leads to useful simplifications in the evaluation of matrix elements involving wave functions for nuclear motion.

The perturbation term in (3.5) can therefore be expanded more usefully in terms of the normal coordinates, $Q_l$ (notice that the transformation between the cartesian displacement coordinates and the $Q_l$ is not an orthogonal transformation):

$$\sum_{l=1}^{3N} Q_l V_l \tag{3.6}$$

The $V_l$ are functions of the electron coordinates, and the position coordinates of the nuclei in their equilibrium configuration: that is, they are formally given by the derivatives

$$\left( \frac{\partial \hat{H}}{\partial Q_l} \right)_{Q=0}.$$

In practice, these coefficients are determined by substituting the values of $(\nabla_q{}^i \hat{H})_{q=0}$ in (3.5), and then expressing the components of the $\mathbf{q}^i$ in terms of the $Q_l$ by means of the transformation matrix which relates the two sets of coordinates. But, since the uniform translation or rotation of the molecule does not change the electronic energy, the coefficients $V_l$ vanish if $l = 1, 2, 3, 4, 5$ or $6$. Thus, the sum over $l$ in (3.6) only contains contributions from the vibrational motions of the molecule.

On taking the $\Psi_j(\mathbf{r}, \mathbf{R}_0, \mathbf{s})$ as expansion functions in the application of perturbation theory, and realizing that the perturbation can be written in either of the equivalent forms

$$\sum_{l=7}^{3N} Q_l V_l \qquad \text{or} \qquad \sum_{i=1}^{N} \mathbf{q}^i \cdot (\nabla_{\mathbf{q}^i}\hat{H})_{\mathbf{q}=0},$$

the coefficients $c_{jk}$ are given by either

$$\frac{1}{E_k - E_j} \sum_{i=1}^{N} \mathbf{q}^i \cdot \int \Psi_j(\mathbf{r}, \mathbf{R}_0, \mathbf{s})(\nabla_{\mathbf{q}^i}\hat{H})_{\mathbf{q}=0}\Psi_k(\mathbf{r}, \mathbf{R}_0, \mathbf{s})\, d\tau$$

or

$$\frac{1}{E_k - E_j} \sum_{l=7}^{3N} Q_l \int \Psi_j(\mathbf{r}, \mathbf{R}_0, \mathbf{s}) V_l \Psi_k(\mathbf{r}, \mathbf{R}_0, \mathbf{s})\, d\tau, \qquad (k \neq j),$$

respectively. The advantage of the second expression for these coefficients is that full use can be made of the symmetry properties of the $Q_l$ and $V_l$ ($Q_l$ and $V_l$ possess the same symmetry properties, as the total Hamiltonian must always be totally symmetric). However, the calculation of the matrix elements of $V_l$ usually proceeds through the former expression, which is then reduced to a summation over the normal coordinates after expressing the components of the $\mathbf{q}^i$ in terms of the $Q_l$.

In the present situation, where the effect of the perturbation is small, $c_{00} \sim 1$, $c_{11} \sim 1$, and hence the expansion for $M_{01}$ is given, to a good approximation, by

$$c_{00}c_{11} \int \Psi_0(\mathbf{r}, \mathbf{R}_0, \mathbf{s}) \left(\sum_i \mathbf{r}_i\right) \Psi_1(\mathbf{r}, \mathbf{R}_0, \mathbf{s})\, d\tau$$

$$+ c_{00} \sum_k{}' c_{k1} \int \Psi_0(\mathbf{r}, \mathbf{R}_0, \mathbf{s}) \left(\sum_i \mathbf{r}_i\right) \Psi_k(\mathbf{r}, \mathbf{R}_0, \mathbf{s})\, d\tau$$

$$+ c_{11} \sum_j{}' c_{j0} \int \Psi_j(\mathbf{r}, \mathbf{R}_0, \mathbf{s}) \left(\sum_i \mathbf{r}_i\right) \Psi_1(\mathbf{r}, \mathbf{R}_0, \mathbf{s})\, d\tau + \dots$$

$$\simeq m_{01} + \sum_k{}' c_{k1}m_{0k} + \sum_j{}' c_{j0}m_{j1}.$$

But $m_{01}$ is zero in the present instance, and hence

$$M_{01} = \sum_k{}' c_{k1}m_{0k} + \sum_j{}' c_{j0}m_{j1}$$

$$= \sum_{k,l} \frac{m_{0k}}{E_1 - E_k} Q_l \int \Psi_k(\mathbf{r}, \mathbf{R}_0, \mathbf{s}) V_l \Psi_1(\mathbf{r}, \mathbf{R}_0, \mathbf{s})\, d\tau$$

$$+ \sum_{j,l} \frac{m_{j1}}{E_0 - E_j} Q_l \int \Psi_j(\mathbf{r}, \mathbf{R}_0, \mathbf{s}) V_l \Psi_0(\mathbf{r}, \mathbf{R}_0, \mathbf{s})\, d\tau,$$

thereby allowing $\mathbf{M}$ to be written in the form

$$\sum_{k,l} \frac{m_{0k}(V_l)_{k1}}{E_1 - E_k} \int \Lambda_0^s(\mathbf{R})Q_l\Lambda_1^t(\mathbf{R})\, d\mathbf{R} + \sum_{j,l} \frac{m_{j1}(V_l)_{j0}}{E_0 - E_j} \int \Lambda_0^s(\mathbf{R})Q_l \times$$

$$\times \Lambda_1^t(\mathbf{R})\, d\mathbf{R}. \quad (3.7)$$

It therefore remains to perform the integrations over the nuclear coordinates in (3.7). These integrations are more tractable if it is assumed that there is no change in equilibrium geometry in the excited state, and that the vibrational motion is harmonic. Thus, since vibration–rotation interactions are ignored, $\Lambda_0^s$ is represented, for example, by the product of oscillator functions, $\eta_0^{s_i}$:

$$\eta_0^0(Q_7)\eta_0^0(Q_8) \ldots \eta_0^0(Q_{21})\xi_{\text{rot}}\xi_{\text{tr}}$$

where it is assumed, for simplicity, that no vibrational quanta are excited in the vibrational state associated with the electronic ground state (all superscripts zero). $\xi_{\text{rot}}$, $\xi_{\text{tr}}$ are the parts of the nuclear wave function corresponding to rotational and translational motion.

The integration over the nuclear coordinates in (3.7) therefore involves the evaluation of

$$I = \int \ldots \int \eta_0^0(Q_7) \ldots \eta_0^0(Q_{21}) \cdot Q_l \cdot \eta_1^{t_7}(Q_7)\eta_1^{t_8}(Q_8) \ldots \eta_1^{t_{21}}(Q_{21})\, dQ_7 \ldots dQ_{21}$$

after integrating over the coordinates corresponding to rotational and translational motion. The superscripts, $t_i$, on the excited state oscillator functions, denote the number of quanta excited of each normal mode. Hence,

$$I = \left(\prod_{i \neq l} \int \eta_0^0(Q_i)\eta_1^{t_i}(Q_i)\, dQ_i\right) \int \eta_0^0(Q_l)Q_l\eta_1^{t_l}(Q_l)\, dQ_l.$$

But, under the assumptions made above, each integral in the round brackets vanishes, unless $t_i = 0$, because of the orthogonality of the oscillator functions $\eta$. Similarly, the integral involving $Q_l$ vanishes unless $t_l = 1$, and $Q_l$ corresponds to a vibration with odd parity (only then does the matrix element of $V_l$ remain non-zero). Thus, only those excited states contribute to $\mathbf{M}$ which involve an excitation of one quantum of an ungerade vibrational normal mode, relative to the ground vibrational state.

It must be recognized, of course, that this treatment of the origin of vibronic intensity represents an over-simplification of the problem. Although the basic ideas are sound enough to explain the greatly reduced intensities in "forbidden" transitions some modifications are necessary if the potential energy surfaces for the two electronic states are markedly different. The various overlap integrals between the oscillator functions in the integral $I$ no longer vanish, and the reader is referred to the work of Liehr[6] for further details.

The theory has been applied by Liehr and Ballnausen[9] to octahedral Ti$^{+++}$ and Cu$^{++}$ complexes, and by Koide and Pryce[10] to octahedral Mn$^{++}$ complexes.

This completes the discussion of vibronic intensity, and the crystal field theoretical treatment of $d^2$ complexes is now analysed.

## 3.4.   $d^2$ octahedral complexes

The possible strong field configurations for two $d$-electrons, in an octahedral environment, are $e_g^2$, $e_g t_{2g}$, and $t_{2g}^2$. Under the influence of $\hat{H}_2$ the degeneracies associated with these configurations (6, 24 and 15, respectively) are partially removed, and each configuration gives rise to a number of singlet and triplet states:

$$e_g^2 \qquad {}^1A_{1g}, {}^1E_g, {}^3A_{2g}$$

$$e_g t_{2g} \qquad {}^1T_{1g}, {}^1T_{2g}, {}^3T_{1g}, {}^3T_{2g}$$

$$t_{2g}^2 \qquad {}^1T_{2g}, {}^1A_{1g}, {}^1E_g, {}^3T_{1g}$$

In the free $V^{+++}$ ion, for example, the ${}^3F$ term forms the ground state, and this is assumed to arise solely from the $3d^2$ configuration: the lowest energy singlet term, ${}^1D$, lies about 0·22 aJ higher in energy.[11] Even in the strong field limit, the triplet state ${}^3T_{1g}$ (predominant configuration $t_{2g}^2$) remains the state of lowest energy, and correlates smoothly with the lower energy atomic triplet states in the limit where the crystal field strength tends to zero: that is, there is a smooth transition to the weak field description.

The understanding of this correlation of strong and weak field configurational wave functions is important, and it is worth while to carry through the analysis for the particular example of ${}^3T_{1g}(t_{2g}^2)$ already cited above.

The three component wave functions associated with the ${}^3T_{1g}$ state, for given $M_S$ (say unity), are denoted by[3]

$$\Psi_{0u} \equiv \Psi_0[{}^3T_{1g}(t_{2g}^2); u, M_S = 1] = N\,|\zeta(1)\eta(2)|,$$

$$\Psi_{0v} \equiv \Psi_0[{}^3T_{1g}(t_{2g}^2); v, M_S = 1] = N\,|\xi(1)\zeta(2)|,$$

$$\Psi_{0w} \equiv \Psi_0[{}^3T_{1g}(t_{2g}^2); w, M_S = 1] = N\,|\eta(1)\xi(2)|.$$

The transformation properties of these determinantal wave functions are readily verified by operating with the group generators. For example, $\Psi_{0u}$ is transformed into $\Psi_{0v}$ after applying the generator $\hat{C}_3^{xyz}$:

$$\hat{C}_3^{xyz}\Psi_{0u} = N\,|\hat{C}_3^{xyz}\zeta(1)\hat{C}_3^{xyz}\eta(2)| = N\,|\xi(1)\zeta(2)| = \Psi_{0v}.$$

The transformation properties of the remaining component wave functions follow in an analogous way, and the overall result can be expressed in the form

$$\hat{C}_3^{xyz}\,\overbrace{\Psi_{0u}\Psi_{0v}\Psi_{0w}} = \overbrace{\Psi_{0u}\Psi_{0v}\Psi_{0w}}\begin{pmatrix} 0 & 0 & 1 \\ 1 & 0 & 0 \\ 0 & 1 & 0 \end{pmatrix}$$

thus yielding a character of zero for the generator $\hat{C}_3^{xyz}$. The remaining generators $\hat{C}_4^z$ and $\hat{J}$ give rise to characters 1 and 3, respectively, and hence the set of three component wave functions span the irreducible representation $T_{1g}$ (see Appendix II).

Even in the absence of the crystal field, $\hat{V}$, the three wave functions, $\Psi_{0\alpha}$, will not be eigenfunctions of $\hat{L}^2$ and $\hat{L}_z$, as the basic one-electron functions are adapted to the octahedral problem where the orbital angular momentum is no

E

longer a constant of the motion: for example,

$$\hat{L}_z\Psi_{0u} = [\hat{l}_z(1) + \hat{l}_z(2)]\Psi_{0u} = N\,|\hat{l}_z(1)\zeta(1)\eta(2)| + N\,|\zeta(1)\hat{l}_z(2)\eta(2)|.$$

Before operating with $\hat{l}_z$ it is first necessary to express the strong field atomic functions in terms of the weak field atomic functions:

$$\zeta = -(i/\sqrt{2})(d_2 - d_{-2}); \qquad \epsilon = (1/\sqrt{2})(d_2 + d_{-2}),$$
$$\eta = -(1/\sqrt{2})(d_1 - d_{-1}); \qquad \xi = (i/\sqrt{2})(d_1 + d_{-1}),$$

where it should be noted that the phase convention of Condon and Shortley[12] is used in defining the complex functions $d_m$. This involves multiplying the angular part of the $3d$ orbitals, $P_2^{|m|}(\cos\theta)\,e^{im\phi}$, by the phase factor $(-1)^{[m+|m|]/2}$, so that the results of the theory of angular momentum can be used directly.

Now, $\hat{l}_z d_m = m d_m$, and hence it follows that

$$\hat{L}_z\Psi_{0u} = -2Ni\,|\epsilon(1)\eta(2)| - Ni\,|\zeta(1)\xi(2)|$$

which is not a constant multiple of $\Psi_{0u}$. Thus the component wave functions of the $^3T_{1g}$ term, in the strong field representation, give rise to a mixture of different free ion term wave functions even in the absence of the crystal field. It is therefore of some interest to discover which atomic terms (triplets) are correlated with the $^3T_{1g}$ term derived from the configuration $t_{2g}^2$.

Consider the component wave function $\Psi_{0w}$; on substituting the appropriate combinations of weak field functions for $\eta$ and $\xi$ this wave function becomes

$$-(i/2)\,|(d_1 - d_{-1})(1)(d_1 + d_{-1})(2)| = -i\,|d_1(1)d_{-1}(2)|$$

which is seen to be a component wave function with $M_L = 0$ and $M_S = 1$. But, in fact, this wave function is a mixture of the $M_L = 0$, $M_S = 1$ component wave functions of the $^3F$ and $^3P$ terms arising from the $d^2$ electron configuration. This is readily seen by noting that, for two $d$-electrons, each with $m_s = \frac{1}{2}$, the maximum permissible value of $L$ is three, and hence the wave function for the $M_L = 3$, $M_S = 1$ component of the $^3F$ term is given directly by

$$|d_2(1)d_1(2)|.$$

The $M_L = 0$, $M_S = 1$ component wave function of this term is obtained by three applications of the step-down operator $\hat{L}_-$. The use of this operator has already been discussed in Vol. 1 but, as a reminder, some of its properties are now restated:

$$\hat{L}_- = \hat{l}_-(1) + \hat{l}_-(2) + \ldots$$
$$\hat{l}_-(1)\phi_{lm_l} = \{(l + m_l)(l - m_l + 1)\}^{1/2}\phi_{lm_l-1}$$
$$\hat{L}_-\Psi(LM_L) = \{(L + M_L)(L - M_L + 1)\}^{1/2}\Psi(LM_L - 1)$$

Hence,

$$\hat{L}_-\Psi[^3F(d^2); M_L = 3, M_S = 1] = 2\,|d_1(1)d_1(2)\,| + \sqrt{6}\,|d_2(1)d_0(2)\,|$$
$$= \sqrt{6}\,|d_2(1)d_0(2)\,|$$
$$= \sqrt{6}.\Psi[^3F(d^2); M_L = 2, M_S = 1]$$

which, after a further application of $\hat{L}_-$,

$$\hat{L}_-\Psi[^3F(d^2); M_L = 2, M_S = 1] = 2\,|d_1(1)d_0(2)\,| + \sqrt{6}\,|d_2(1)d_{-1}(\,2)\,|$$
$$= \sqrt{10}.\Psi[^3F(d^2); M_L = 1, M_S = 1]$$

identifies $\Psi[^3F(d^2); M_L = 1, M_S = 1]$ as

$$(1/\sqrt{5})\{\sqrt{2}\,|d_1(1)d_0(2)\,| + \sqrt{3}\,|d_2(1)d_{-1}(2)\,|\}.$$

A final application of $\hat{L}_-$ then yields

$$\Psi[^3F(d^2); M_L = 0, M_S = 1] = (1/\sqrt{5})\{2\,|d_1(1)d_{-1}(2)\,| + |d_2(1)d_{-2}(2)\,|\}. \quad (3.8)$$

The $M_L = 0$, $M_S = 1$ component wave function of the $^3P$ term (the only other triplet term arising from the configuration $d^2$) must be orthogonal to (3.8), and this is sufficient to determine its form as

$$\Psi[^3P(d^2); M_L = 0, M_S = 1] = (1/\sqrt{5})\{2\,|d_2(1)d_{-2}(2)\,| - |d_1(1)d_{-1}(2)\,|\}. \quad (3.9)$$

Hence, $\Psi_{0w}$ reduces to the following combination of the $M_L = 0$, $M_S = 1$ component wave functions of the $^3F$ and $^3P$ atomic terms:

$$2/\sqrt{5}.\Psi(^3F) - 1/\sqrt{5}.\Psi(^3P) \quad (3.10)$$

with similar results for $\Psi_{0u}$ and $\Psi_{0v}$.

The correlation of the molecular $^3T_{1g}(t_{2g}^2)$ term with the $^3P$ and $^3F$ term wave functions of the free ion illustrates very clearly how the octahedral field destroys the constancy of the orbital angular momentum, since free ion terms with $L = 1$ and $L = 3$ are mixed together under the influence of $\hat{V}$.

The complete correlation of strong field configurations with weak field term wave functions is shown schematically in Fig. 3.2 for the triplet states only.

Figure 3.2 shows that the three lowest energy, spin allowed, transitions will again be of low intensity, as there is no change in parity in all cases. However, some additional bands are now expected in the absorption spectrum on account of the perturbation due to spin-orbit coupling: this leads to a mixing of singlet and triplet states, thereby vitiating the use of $S$ as a good quantum number for classifying electronic states.

$V^{+++}$ is the best example of a transition metal ion with the $d^2$ electron configuration. The absorption spectrum of the hexa-hydrated ion, as measured by Hartman and Furlani,[13] shows two weak absorption bands with maximum intensity at 0·340 aJ and 0·473 aJ. The third band, which is expected to appear at higher energy, is obscured by the presence of an intense (allowed)

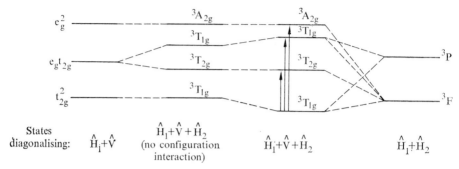

States
diagonalising:    $\hat{H}_i + \hat{V}$    $\hat{H}_i + \hat{V} + \hat{H}_2$    $\hat{H}_i + \hat{V} + \hat{H}_2$    $\hat{H}_i + \hat{H}_2$
(no configuration
interaction)

FIG. 3.2. The correlation of strong field configurations with free ion terms for
$d^2$ systems, in an octahedral environment. Possible spin-allowed transitions
are indicated by vertical arrows.

absorption band, which is thought to arise from a charge-transfer transition.
In fact, all three absorption bands have been observed by Low[14] for $V^{+++}$ in
$Al_2O_3$. Low also observed four other weak bands which are thought to be
associated with transitions from the $^3T_{1g}$ ground state to upper states which
are predominantly singlet in character.

The complete verification of these experimental results, on the basis of the
crystal field model, is a complex task—particularly when the transition metal
ion is embedded in a host lattice—and the reader is referred to the work of
Liehr and Ballhausen (see Bibliography) for a treatment of the free complex
ion which includes the effects of spin-orbit coupling, but not vibronic coupling.
In the present discussion, though, both of these effects are neglected in order
to show how the crystal field theory can be parameterized to obtain a value of
$\Delta$ from the experimental results.

The analysis is more involved than in the case of the simple $d^1$ metal ion
complex, because allowance must be made for the configurational mixing
between the two $^3T_{1g}$ electronic states arising from different strong field
configurations. The modified energies of the ground and excited $^3T_{1g}$ states
are obtained as the roots of the $2 \times 2$ secular determinant arising from the
interaction between a given pair of component wave functions—say the $w$
components with $M_S = 1$:

$$\begin{vmatrix} H_{00} - E & H_{01} \\ H_{10} & H_{11} - E \end{vmatrix} = 0,$$

where, in an obvious notation,

$$H_{00} = \iint \Psi_0[^3T_{1g}(t_{2g}^2); w, M_S = 1]\hat{H}\,\Psi_0[^3T_{2g}(t_{2g}^2); w, M_S = 1]\,dr\,ds,$$

$$H_{11} = \iint \Psi_1[^3T_{1g}(t_{2g}e_g); w, M_S = 1]\hat{H}\,\Psi_1[^3T_{2g}(t_{2g}e_g); w, M_S = 1]\,dr\,ds,$$

$$H_{01} = \iint \Psi_0[^3T_{1g}(t_{2g}^2); w, M_S = 1]\hat{H}\,\Psi_1[^3T_{2g}(t_{2g}e_g); w, M_S = 1]\,dr\,ds.$$

The explicit forms of the determinantal wave functions, required in the evaluation of the matrix elements, are best found in chapter 9 and table A24 of Griffith:[3]

$$\Psi_0[^3T_{1g}(t_{2g}^2); w, M_S = 1] = |\eta(1)\xi(2)| = -i|d_1(1)d_{-1}(2)|,$$

$$\Psi_1[^3T_{1g}(t_{2g}e_g); w, M_S = 1] = |\zeta(1)\epsilon(2)| = i|d_2(1)d_{-2}(2)|.$$

On expanding $\hat{H}$ as $\hat{H}_1 + \hat{H}_2 + \hat{V}$, and noting that the sum of the first two operators represents the Hamiltonian for the free ion, the matrix element $H_{00}$ becomes

$$\tfrac{4}{5}E(^3F) + \tfrac{1}{5}E(^3P) + 2\int \eta\hat{V}\eta\,\mathrm{d}\mathbf{r}_1$$

after using (3.10), and noting that

$$\int \xi\hat{V}\xi\,\mathrm{d}\mathbf{r}_1 = \int (\hat{C}_4^z\xi)\hat{C}(_4^z\hat{V})(\hat{C}_4^z\xi)\,\mathrm{d}\mathbf{r}_1 = \int (\hat{C}_4^z\xi)\hat{V}(\hat{C}_4^z\xi)\,\mathrm{d}\mathbf{r}_1 = \int \eta\hat{V}\eta\,\mathrm{d}\mathbf{r}_1.$$

The evaluation of the matrix elements involving $\Psi_1$ proceeds by first expressing $|d_2(1)d_{-2}(2)|$ in terms of the free ion wave functions by using (3.8) and (3.9):

$$|d_2(1)d_{-2}(2)| = 1/\sqrt{5}.\Psi(^3F) + 2/\sqrt{5}.\Psi(^3P),$$

The two remaining matrix elements are therefore given by

$$H_{11} = \tfrac{1}{5}E(^3F) + \tfrac{4}{5}E(^3P) + \int \eta\hat{V}\eta\,\mathrm{d}\mathbf{r}_1 + \int \epsilon\hat{V}\epsilon\,\mathrm{d}\mathbf{r}_1$$

$$H_{01} = -\tfrac{2}{5}[E(^3F) - E(^3P)] = \tfrac{2}{5}x$$

and the roots of the secular determinant are found to be

$$\frac{E(^3F) + E(^3P)}{2} + 2\int \eta\hat{V}\eta\,\mathrm{d}\mathbf{r}_1 + \tfrac{1}{2}\Delta \pm \tfrac{1}{2}\{x^2 + \tfrac{6}{5}x\Delta + \Delta^2\}^{1/2}.$$

One of the $M_S = 1$ components of the $^3T_{2g}(t_{2g}e_g)$ state is given by

$$|\zeta(1)\theta(2)| = -(i/\sqrt{2})\{|d_2(1)d_0(2)| - |d_{-2}(1)d_0(2)|\},$$

which is seen to be an admixture of the $M_L = 2$ and $M_L = -2$ components of the $^3F$ term. The energy of this component wave function of the $^3T_{2g}$ term follows immediately as

$$E(^3T_{2g}) = E(^3F) + 2\int \eta\hat{V}\eta\,\mathrm{d}\mathbf{r}_1 + \Delta.$$

The energy of the $^3A_{2g}(e_g^2)$ term is obtained in a similar way, after first recognizing that the term wave function, in the strong field representation, reduces to a linear combination of the $M_L = \pm 2$, $M_S = 1$ components of the atomic $^3F$ term wave functions:

$$|\epsilon(1)\theta(2)| = (1/\sqrt{2})\{|d_2(1)d_0(2)| + |d_{-2}(1)d_0(2)|\}$$

with energy

$$E(^3A_{2g}) = E(^3F) + 2 \int \eta \hat{V} \eta \; \mathbf{dr}_1 + 2\Delta.$$

The transition energies for the transitions $^3T_{2g} \leftarrow {}^3T_{1g}$, $^3T_{1g} \leftarrow {}^3T_{1g}$, and $^3A_{2g} \leftarrow {}^3T_{1g}$ are therefore given by

$$-\tfrac{1}{2}x + \tfrac{1}{2}\Delta + \tfrac{1}{2}\{x^2 + \tfrac{6}{5}x\Delta + \Delta^2\}^{1/2}$$

$$\{x^2 + \tfrac{6}{5}x\Delta + \Delta^2\}^{1/2}$$

and

$$-\tfrac{1}{2}x + \tfrac{3}{2}\Delta + \tfrac{1}{2}\{x^2 + \tfrac{6}{5}x\Delta + \Delta^2\}^{1/2},$$

respectively.

The problem remains of finding the best values of $\Delta$ and $x$ which lead to transition energies which are in accord with the experimental data. As only two absorption bands have been observed for the hydrated $V^{+++}$ ion in solution, it seems sensible, therefore, to find the values of the two parameters which will fit these results. This is a simple arithmetical problem, and leads to $\Delta = 0 \cdot 4101$ aJ and $x = 0 \cdot 2316$ aJ. The value of $x$ should be compared with the value of $0 \cdot 2567$ aJ for the experimental separation in energy between the $^3F$ and $^3P$ terms in the free ion (the number actually given is the difference in average energies, because the effects of spin-orbit coupling give rise to small term splittings).

The energy of the $^3A_{2g}$ state is predicted to lie $0 \cdot 41$ aJ above that of the $^3T_{2g}$ state, leading to an energy of about $0 \cdot 75$ aJ for the transition $^3A_{2g} \leftarrow {}^3T_{1g}$. This transition, which lies in the region of strong absorption, is difficult to characterize in the solution spectrum. However, the absorption spectrum of $V^{+++}$, embedded in $Al_2O_3$, displays three absorption bands at $0 \cdot 345$ aJ, $0 \cdot 500$ aJ and $0 \cdot 685$ aJ which Low[14] associates with the three transitions under discussion. Although the crystal fields are different in the two systems, because the nearest-neighbour species are now $O^{--}$ ions, rather than water molecules, it is gratifying to find experimental evidence for a $d$–$d$ absorption band in the region of $0 \cdot 69$ aJ.

The crystal field theoretic approach has, as a central feature, the assumption that the free ion $d$-orbitals can be used for describing the electron distribution around the metal ion in the complex. As shown, this leads to a favourable situation in $d^2$ complexes, for example, as the three transition energies are determined solely by the parameters $x$ and $\Delta$.† On the other hand, if the radial functions for the $t_{2g}$ and $e_g$ orbitals are different, then the relative energies of the excited triplet states are determined by eleven parameters (nine parameters arising from the two-electron integrals over radial functions, $(\gamma - \epsilon)$, and $\int \eta \hat{V} \eta \; \mathbf{dr}_1 - \int \theta \hat{V} \theta \; \mathbf{dr}_1$), thereby making it virtually impossible to find a unique fit of the theoretical energy differences to

---

† Some authors prefer to express state energies in terms of the Racah parameters $A$, $B$ and $C$, which are given as sums of two-electron integrals involving the $3d$ radial functions. In the present example, the parameter $x$ can be identified with $15B$.

the experimental data. This effect is important if the central metal ion is in a low oxidation state, as the $3d$ orbitals are then more susceptible to the perturbing influence of the ligands.

The extension of the crystal field theory to deal with other $d$-electron configurations is straightforward in principle, but the pattern of energy levels becomes much more complex. As a consequence, there is a distinct possibility of ambiguity in the assignment of spectral absorption bands, and the reader is referred to the standard texts on ligand field theory for a detailed discussion (see Bibliography).

In any event, the general method of assigning spectral bands is based on the parameterization of the calculations in order to fit the experimental results. This procedure is, in itself, of course, open to criticism, but it is anticipated that, for a given transition metal ion, the changes in the parameters for different ligands will give useful information on the changes in metal ion–ligand interactions. When the theory is used in this form it is therefore a phenomenological theory, and not too much reliance can be placed on the absolute values of a given set of parameters.

The results obtained from the application of simple crystal field theory should therefore be regarded with some care, as several factors add to the uncertainty in ascribing changes in the empirical parameters to observed physical properties. For example, ligand–ligand interactions are invariably neglected, and this may have important implications when trying to rationalize crystal field parameters involving the same ligand coordinated to different transition metal ions. The effects of electron delocalization are also completely ignored, although there have been some attempts to cope with this problem (see Bibliography). Finally, any change in shape or size of the central metal ion's atomic orbitals is generally ignored, for reasons already discussed above.

As the simple theory explicitly excludes detailed consideration of polarization and charge transfer, these effects are all ascribed, in an unknown way, to the basic parameters of the theory—which, for example, are $x$ and $\Delta$ in the case of $d^2$ complexes. This is why it is difficult to attach much significance to variations in these parameters for different complexes. Although the polarization of the metal ion can be treated within the crystal field framework, by allowing for the configuration interaction between terms arising from the occupancy of $4s$, $4p$, ... metal ion atomic orbitals, there is still no really satisfactory method of dealing with the effects of charge transfer. Nevertheless, as discussed in Chapter 1, the expansion of the basic set of atomic orbitals on the metal ion goes some way, in principle, towards describing the effects of charge transfer. But this approach is unlikely to be satisfactory within the crystal field formalism, as the orbital structure of the ligands is completely disregarded.

An appraisal of the crystal field approach is given in Section 3.5. This is then followed by an examination of the methods used for circumventing the problem of treating the ligands as point charges.

## 3.5.    An appraisal of crystal field theory

Although the crystal field theory appears to yield a useful method for discussing the electronic structure of transition metal ion complexes, the quantitative nature of the information is illusory. Successful applications of the theory are usually found only for highly symmetrical ligand groupings: a situation which is highlighted by the distinct lack of success with complex ions possessing less symmetrical dispositions of ligands.

There are several deficiencies of the model. First, only atomic orbitals of the central metal ion are explicitly considered. Secondly, overlap and exchange interactions with the ligands are neglected. On the basis of these deficiencies alone, it is perhaps surprising that the theory yields any sensible results at all, a feature which arises only through a fortuitous cancellation of errors in the parameterization procedure. The origin of these errors can be analysed by considering the complete wave function for the complex ion, within the set of assumptions normally made in the crystal field approach.

The simulation of the ligands by suitably chosen sources of Coulombic potential implies the use of the non-antisymmetrized wave function

$$\Psi_0 = \phi_M(1, 2, \ldots, n)\phi_{L1}\phi_{L2}, \ldots, \phi_{L6} \qquad (3.11)$$

corresponding, in effect, to a valence bond ionic structure. $\phi_{Li}$ is the many-electron wave function associated with ligand $Li$: for example, a $F^-$ ion or an $H_2O$ molecule, and $\phi_M$ is the antisymmetrized metal group function, describing the electronic state of the $n$, $d$ electrons.

Thus, in the crystal field model, excited states of the complex ion are represented by product wave functions of the form (3.11):

$$\Psi_1 = \phi_{M'}(1, 2, \ldots, n)\phi_{L1}\phi_{L2}, \ldots, \phi_{L6}, \qquad (3.12)$$

where $\phi_{M'}$ differs from $\phi_M$ only in the occupancy of the basic strong field orbitals.

The Hamiltonian for the complex ion may be written as

$$\mathcal{H} = -\frac{1}{2}\sum_m \nabla_m^2 - \frac{1}{2}\sum_{\alpha, i}\nabla_{\alpha(i)}^2 - \sum_{\alpha, m}\frac{Z_\alpha}{r_{\alpha m}} - \sum_{\alpha, i}\frac{Z_M}{r_{M\alpha(i)}} - \sum_m\frac{Z_M}{r_{Mm}}$$

$$- \sum_{\substack{\alpha, \alpha' \\ i}}\frac{Z_\alpha}{r_{\alpha\alpha'(i)}} + \sum_{m<m'}\frac{1}{r_{mm'}} + \frac{1}{2}\sum_{\substack{\alpha, \alpha' \\ i, j \\ (i \neq j)}}\frac{1}{r_{\alpha(i)\alpha'(j)}} + \sum_{\alpha, i, m}\frac{1}{r_{m\alpha(i)}} + NN,$$

where, for simplicity, the effects of spin-orbit coupling are ignored. The $m$ labels electrons on the metal ion; $\alpha(i)$ labels electron $i$ on ligand

$\alpha(\alpha = L1, L2, \ldots, L6)$; $M$ labels the metal ion nucleus; $Z_M$ is the effective nuclear charge of the metal ion, and $NN$ is the energy arising from the nuclear–nuclear repulsions.

Hence, the energy of the ground state, which is assumed to be given by (3.11), takes the form

$$E_0 = \int \Psi_0 \hat{H} \Psi_0 \, d\tau$$

$$= \int \phi_M \left[ -\frac{1}{2} \sum_m \nabla_m^2 - \sum_m \frac{Z_M}{r_{Mm}} + \frac{1}{2} \sum_{m,m'}' \frac{1}{r_{mm'}} \right] \phi_M \, d\tau_M$$

$$+ \int \phi_M \left[ \sum_\alpha \int \phi_{L1} \ldots \phi_{L6} \left[ \sum_m \left\{ \sum_i \frac{1}{r_{m\alpha(i)}} - \frac{Z_\alpha}{r_{\alpha m}} \right\} - \sum_i \frac{Z_M}{r_{M\alpha(i)}} \right] \times \right.$$

$$\left. \times \phi_{L1}, \ldots, \phi_{L6} \, d\tau_L \right] \phi_M \, d\tau_M$$

$$+ \int \phi_{L1} \ldots \phi_{L6} \left[ \frac{1}{2} \sum_{\substack{\alpha,\alpha' \\ i,j(i \neq j)}} \frac{1}{r_{\alpha(i)\alpha'(j)}} - \frac{1}{2} \sum_{\alpha,i} \nabla_{\alpha(i)}^2 - \sum_{\alpha,\alpha'} \frac{Z_\alpha}{r_{\alpha\alpha'(i)}} \right] \times$$

$$\times \phi_{L1}, \ldots, \phi_{L6} \, d\tau_L + NN$$

$$= E_M + E_{ML} + E_L$$

where $E_M$, $E_{ML}$ and $E_L$ denote the contributions to the molecular energy associated with the transition metal ion, the metal ion–ligand interactions, and the ligand–ligand interactions, respectively.

The second term, $E_{ML}$, which is of particular interest, represents the net Coulomb interaction between the metal ion and the ligands. And, on writing each ligand wave function as a product of $N/2$ doubly occupied one-electron functions, $\psi_{\alpha n}$ (there is no need to antisymmetrize as all exchange integrals are neglected), this term becomes

$$\int \phi_M \left[ \sum_{\alpha,m} \left\{ -\frac{Z_\alpha}{r_{\alpha m}} + 2 \sum_n \int \frac{\psi_{\alpha n}(l)\psi_{\alpha n}(l)}{r_{lm}} \, d\tau_l \right\} \right.$$

$$\left. - \sum_{\alpha,n} 2Z_M \int \frac{\psi_{\alpha n}(l)\psi_{\alpha n}(l)}{r_{lM}} \, d\tau_l \right] \phi_M \, d\tau_M, \quad (3.13)$$

where $l$ labels electrons associated with ligand $\alpha$, and $n = 1, 2, \ldots, N/2$. The terms in the inner brackets represent the potential, at the position of electron $m$, arising from the nucleus and electrons associated with ligand $\alpha$.

Now, in the crystal field model, the metal–ligand charge distributions are assumed to be non-penetrating: this ensures that the ligand atomic orbitals are not involved explicitly in determining the electronic structure of the complex ion. In this situation, the potential due to each ligand is the same as that arising from the total ligand electronic charge, concentrated on its respective nucleus, plus the potential due to each nuclear charge.

The values of the integrals in (3.13), describing the interaction between the metal ion and the ligands, must therefore be given their point charge values, in order to be consistent with the assumptions of the crystal field model:

$$\int \frac{\psi_{\alpha n}(l)\psi_{\alpha n}(l)}{r_{lm}} \Rightarrow \frac{1}{r_{\alpha m}}$$

$$\int \frac{\psi_{\alpha n}(l)\psi_{\alpha n}(l)}{r_{lM}} \Rightarrow \frac{1}{r_{\alpha M}}$$

The interaction energy, $E_{ML}$, therefore reduces to

$$\int \phi_M \left[ \sum_{\alpha, m} \frac{(N - Z_\alpha)}{r_{\alpha m}} \right] \phi_M \, d\tau_M + x_L$$

$$= \int \phi_M \left[ \sum_{\alpha, m} - Q/r_{\alpha m} \right] \phi_M \, d\tau_M + x_L$$

$$= \int \phi_M \hat{V} \phi_M \, d\tau_M + x_L,$$

where $Q$, equal to $Z_\alpha - N$, is the effective ligand charge, in units of $e$ (see p. 58), and $x_L$ denotes the terms which are independent of metal electron coordinates.

Hence, on taking $E_L + x_L$ as the zero of energy, the expression (3.14), for the energy of the metal ion $d$ electrons in the field of the ligands, becomes identical with the one used in Section 3.2.

This discussion shows clearly that there are two main prerequisites in treating the ligands as point charges. First, the full effects of antisymmetry are neglected, in so far as permutations of electrons between metal ion and ligands are ignored; secondly, the various two-centre integrals in the energy expression are replaced by their classical point-charge analogues, so obviating the need for defining explicit forms for the basic atomic orbitals. Thus, both of these assumptions effectively limit the applicability of crystal field theory to systems in which the metal ion–ligand interactions are weak; in practical terms, this corresponds to having small metal–ligand overlap integrals as well as a fairly large difference in the metal and ligand valence orbital energies. Under these conditions, it may be possible to represent the metal ion–ligand interactions in terms of an electrostatic perturbation, since the non-classical contributions to the energy, arising through overlap and exchange effects, will be much less important. However, unless it is possible to assess the contribution of these non-classical terms, it always remains doubtful whether the crystal field model is adequate. For this reason, there have been a number of attempts to generalize the formalism of crystal field theory in order to allow for orbital structure on the ligands.

The assumption of replacing two-centre integrals by their classical point charge analogues can be transcended by evaluating the integrals in (3.13), for a given choice of metal ion and ligand atomic orbitals; in this approach, the ligand potentials (the terms in curly brackets in (3.13)) depend upon the analytic forms, and occupation numbers, of the ligand atomic orbitals. This seems a more attractive proposition than the one of simply simulating the ligands by suitably chosen point charges. Unfortunately, any attempt to improve the calculation by this method fails because, as shown by Kleiner,[15] the use of "proper" ligand potentials results in the preferential stabilization of the metal ion $e_g$ orbitals, thereby leading to a negative value for $\Delta$. This effect arises through the dominating influence of the potential due to the ligand nuclear charges $Z_\alpha$, and is at variance with the choice of an effective negative charge required for correlating the observed spectral transitions. In fact, this apparent paradox arises through the neglect of exchange and overlap terms in the energy expression; that is, through the incomplete treatment of the effects of antisymmetry. By neglecting these effects, the metal ion $d$ electron is allowed to penetrate the electron distribution around the ligands far too readily, and hence experiences a potential which is excessively attractive in the vicinity of the ligands. Both Tanabe and Sugano[16] and Phillips[17] have shown that a proper treatment of the effects of orbital non-orthogonality leads to a value of $\Delta$ with the correct sign.

It is now apparent that, although the simple crystal field model predicts the right sense of the $d$-orbital splitting, the basic description of the electronic structure is deficient in many respects; in particular, there is no reason to suppose that the strong field $t_{2g}$ and $e_g$ orbitals can be adequately described as pure metal ion $3d$ atomic orbitals. As already seen in Chapter 2, the $e_g$ orbitals are much more likely to be delocalized over the molecular framework, in view of their favourable orientation for interacting with $\sigma$-bonding ligands. Consequently, it will not be possible, in general, to discuss objectively any variation in $\Delta$, with ligand type, as $\Delta$ is itself a complicated function of the metal ion and ligand electronic structure (the reader is referred to the work of Jørgensen[18] for an alternative view in which $\Delta$ is expressed in the form $\Delta = fg$, where $f$ and $g$ are functions depending upon metal ion and ligand electronic structure, respectively). The electronic structure of transition metal ion complexes must therefore be discussed in a manner which explicitly recognizes the electronic structure of the ligands from the outset; the problem then reduces to one of finding the best way of incorporating the effects of ligand orbital structure into the theory. There have been a number of attempts to develop satisfactory theories, but they all suffer from limitations arising from theoretical or computational difficulties inherent in their formulation; the most common one being the ubiquitous problem of metal–ligand atomic orbital non-orthogonality.

Tanabe and Sugano[16] have used a simple valence bond model to calculate the value of $\Delta$ in $[\mathrm{Cr(H_2O)_6}]^{+++}$. They used wave functions like (3.11) and (3.12), but fully antisymmetrized, to describe the ground and excited states of the complex, respectively. The appropriate metal group functions, $\phi_M$ and $\phi_{M'}$, corresponded to the $^4A_{2g}$ and $^4T_{2g}$ states arising from the $d$-electron configurations $t_{2g}^3$ and $t_{2g}^2 e_g^1$, respectively. With this choice of excited state wave function, the difference in total energies is just equal to $\Delta$ (that is, $\Delta$ would be the separation between these two states if crystal field theory were used). The Tanabe–Sugano calculation has the advantage that the metal–ligand orbital non-orthogonality can be handled relatively easily, as both electronic wave functions are represented by single determinants. However, the neglect of covalency effects, as described by contributions from structures based on $Cr^{++}$, or $Cr^{+}$, is a deficiency of the calculation which is difficult to overcome because of (a) the non-orthogonality of the contributing structures, and (b) the difficulty in choosing the best forms for the $d$-orbitals on the central metal ion. Both of these difficulties clearly limit the use of valence bond theory for obtaining an adequate description of covalency, or charge transfer, effects in complex ions. For this reason, a form of molecular orbital theory is often considered as a more practical alternative; but, even in this model, there are still serious problems to be overcome before satisfactory results can be obtained, as will become apparent in Section 3.7.

The main advantage in using molecular orbital theory is that it is possible to allow for electron delocalization, between metal ion and ligands, from the beginning of the calculation. This is an important consideration, as there is a growing body of experimental evidence which indicates a more active participation by the ligands in the complex ion. For example, nuclear magnetic resonance studies on crystalline $KNiF_3$, in which the unit of interest is the $[NiF_6]^{----}$ octahedron, indicate that some spin density is transferred to the ligands.[19] Also, in complexes like $[V(H_2O)_6]^{+++}$, very intense spectral bands are observed to higher frequency from the $d$–$d$ bands; and these are usually ascribed to transitions involving a transfer of charge between the metal ion and the ligands.

## 3.6.  Molecular orbital theory of complex ions

In the molecular orbital approach, the basic one-electron orbitals, previously centred on the transition metal ion, are now allowed to encompass both the metal ion and the ligands. Thus, within the LCAO approximation, the molecular orbitals are written in the form (see (1.43))

$$\psi_i = \sum_\lambda c^0_{\lambda i}\chi_\lambda = \sum_m c^0_{mi}\chi_m + \sum_{\alpha, n} c^0_{\alpha n}\chi_{\alpha n}$$

where the $\chi_k$ are orthogonalized atomic orbitals, and the coefficients, $c^0_{ki}$, are determined by energy minimization.

If the complex ion has a closed shell structure, then the equations determining the $c^0_{ki}$ are given by (1.44). In this situation, with $\hat{h}$ possessing the total symmetry of the molecular point group (see (1.35), where the matrix $O$ can be taken as the matrix representative of a symmetry operation), the molecular orbitals can be labelled by the irreducible representations of $G$. For a highly symmetrical molecule, many of the coefficients, $c_{ki}^0$, are determined by symmetry, and it is often preferable to make an expansion of the molecular orbitals in terms of symmetry orbitals, rather than ordinary atomic orbitals; that is, the set of ordinary atomic orbitals is replaced by the same number of linearly independent symmetry orbitals, which are found by repeated application of the operator

$$\sum_{\hat{R}} \chi_\Gamma^*(\hat{R})\hat{R}$$

as already described in Chapter 2. The various permitted irreducible representations, $\Gamma$, are first found by resolving the reducible representation in the usual way.

The symmetry adapted combinations of atomic orbitals, appropriate to $O_h$, are given in Appendix III, where the local axes on the ligand atoms are parallel to the reference axis system on the central metal ion as used in Fig. 2.1. The molecular orbitals are now expressed in the form

$$\psi_j = \sum_S d_{Sj}(\Gamma)\lambda_S(\Gamma),$$

where $S$ labels metal ion and ligand symmetry adapted combinations of atomic orbitals, and it only remains to evaluate the matrix elements of $\hat{h}$; this requires explicit forms for all the atomic orbitals appearing in the LCAO expansion, unless the matrix elements are being estimated semi-empirically. The choice of metal ion atomic orbitals presents the main problem, and there is a tendency to use orbitals appropriate to prejudged metal ion oxidation states in the molecule. This kind of choice is often one of necessity, because it is too time consuming to optimize the molecular energy with respect to each orbital exponent. In addition, if the molecular orbitals are constructed out of valence atomic orbitals only, then it is necessary to orthogonalize these orbitals to all atomic core orbitals.

The results of a typical molecular orbital calculation depend upon the choice of basic atomic orbitals; the method used for calculating integrals; and the nature, and number, of ligands.

In the case of an octahedral (closed shell) transition metal ion complex, involving only central ion $3d$, $4s$ and $4p$ atomic orbitals, and ligand $\sigma$ hybrid orbitals, the pattern of energy levels is expected to be of the form shown in Fig. 3.3.

In this model, the interaction between each pair of symmetry orbitals, transforming like the same component of the same irreducible representation, results in bonding and antibonding molecular orbitals; the separation in energy between which is determined by the magnitude of the matrix element of $\hat{h}$ between the two appropriate symmetry orbitals. The core atomic orbitals have been excluded from the discussion but, clearly, they must be considered in a more complete treatment of the problem.

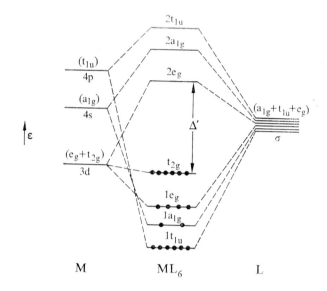

FIG. 3.3. A schematic molecular orbital energy diagram for a closed shell ($d^6$) octahedral complex. The effects of $\pi$-bonding are neglected.

The separation in energy between the highest filled and lowest unfilled molecular orbital is formally equivalent to the parameter $\Delta$ of crystal field theory; but there is no question of $\Delta'$ being identified with $\Delta$, as the usual form of crystal field theory requires the $e_g$ and $t_{2g}$ one-electron orbitals to be pure metal ion $3d$ orbitals (with the same radial functions), in order to limit the number of parameters appearing in the theory. It should also be noticed that sensible results are not obtained from the molecular orbital model unless $4s$ and $4p$ atomic orbitals are included in the LCAO expansion of the molecular orbitals. In crystal field theory, this would correspond to including terms arising from the excitation of one, or more, electrons from the $3d$ shell of the metal ion to the $4s$ or $4p$ shell; but this is rarely, if ever, allowed for in the application of crystal field theory, as the number of parameters becomes too unwieldy.

It is also clear that the unoccupied $e_g$ molecular orbital cannot be described as a mono-centric one-electron orbital, as it contains significant contributions from ligand atomic orbitals. Thus, any comparison with crystal field theory must be made through the more general formalism, which admits the $e_g$ and $t_{2g}$ orbitals to have different radial functions. As this is not easy, it is simpler to perform the molecular orbital calculation instead, and merely interpret the strong field orbitals as molecular orbitals. This approach has the advantage that it automatically includes contributions from covalent and ionic structures, although not with optimum weights; and this is probably more significant than the results of a generalized crystal field calculation which is, in reality, still only a valence bond calculation of limited scope.

The pattern of one-electron orbital energies, obtained from a single configuration molecular orbital calculation, is often useful in qualitative discussions for predicting the number, and nature, of possible electronic transitions. Of course, if there is only one allowed transition, out of several possible transitions, then there is usually no difficulty in assigning the band in question. In the more general situation, however, where there are usually several absorption bands to assign, the use of an orbital energy diagram may lead to misleading predictions—particularly if any of the molecular orbitals involved in the transition are degenerate. In addition, as already seen in Section 1.5, the difference in energy between the ground state and an excited state is not equal to the appropriate difference in orbital energies; there are also contributions from Coulomb and exchange integrals involving the molecular orbitals with occupation number unity.

The only justification for using orbital energy differences, for estimating differences in transition energies, therefore rests on the assumption that these extra two-electron integrals give the same contribution to each excited state of interest. This is unlikely to be a reasonable assumption for heteropolar molecules, as there is usually a significant change in charge distribution after exciting an electron into an antibonding molecular orbital: the sums of Coulomb and exchange integrals will therefore be different for each excited state. On the other hand, if there is little change in polarity, but the orbitals involved in the transition are degenerate, then orbital energy differences are still of little use, as the degeneracy of the states arising from the single electron excitation is removed by $\hat{H}_2$. The $MnO_4^-$ ion provides an example of this situation, as described shortly.

An interesting example, which illustrates the importance of the extra Coulomb and exchange integrals referred to in the last paragraph, is found in the recent calculations of Peyerimhoff[20] on the formate anion, $HCOO^-$ (not strictly an inorganic molecule, but the example will make its point). In this molecule ion, the three one-electron excitations $2b_1 \leftarrow 1a_2$, $2b_1 \leftarrow 6a_1$ and $2b_1 \leftarrow 4b_2$ give rise to singlet excitation energies of 1·68 aJ, 1·23 aJ and 1·22 aJ, respectively. However, the corresponding differences in orbital energies of 2·47 aJ, 2·67 aJ and 2·69 aJ, respectively, are much larger in magnitude, and follow the reverse order of the excitation energies.

The procedure of utilizing the results of a single configuration molecular orbital wave function, for interpreting spectral properties of complex ions, will only be useful if both the ground and excited states of interest are adequately represented by single configurational wave functions; that is, the effects of configuration interaction are unimportant. But this will only be approximately true if there is a sufficiently strong metal–ligand interaction, in order to make $\Delta'$ as large as possible. In this situation, the closed-shell ground state configuration . . . $t_{2g}^6$, for example, gives rise to a $^1A_{1g}$ molecular term, and excited states can be constructed from configurations in which an electron has been excited from the $t_{2g}$ to the antibonding $e_g$ molecular orbital. This description is adequate for most complexes of $Co^{+++}$, and some complexes of $Fe^{++}$ where, for example, in $[Fe(CN)_6]^{----}$ or $[Co(H_2O)_6]^{++}$, the dominant configuration . . . $t_{2g}^6$ gives rise to the ground $^1A_{1g}$ term. But $^1A_{1g}$ terms also arise from the excited configurations . . . $t_{2g}^4 e_g^2$ and . . . $t_{2g}^2 e_g^4$, and it may be necessary to improve the quality of the ground-state wave function by allowing for the configuration interaction between the three approximate $^1A_{1g}$ wave functions. Similarly, the $^1T_{1g}$ and $^1T_{2g}$ excited states, arising from the singly excited configuration $t_{2g}^5 e_g^1$, can also interact with wave functions of the same symmetry arising from the excited configurations

$$t_{2g}^4 e_g^2 (^1T_{1g} + 3\,^1T_{2g}), \quad t_{2g}^3 e_g^3 (2\,^1T_{1g} + 2\,^1T_{2g}) \quad \text{and} \quad t_{2g}^2 e_g^4 (^1T_{2g}).$$

It is now obvious that the simplicity of the single configuration molecular orbital theory is lost, and so are any hopes of using the pattern of orbital energies as a basis for rationalizing the energies of the observed spectral transitions.

A good example of this kind of calculation, the results of which also illustrate the importance of configuration interaction, is provided by the recent work of Brown et al.[21] on the electronic structure of the permanganate ion. These authors used a form of semi-empirical self-consistent molecular orbital theory, in which the invariance requirements, described in Chapter 1, are satisfied. The zeroth order $^1A_1$ ground state arises from the valence electron configuration $1t_2^6 1e^4 1a_1^2 2t_2^6 1t_1^6$. The lowest energy $^1T_1$, $^1E$ and $^1T_2$ excited electronic states arise from the single electron excitations $2e \leftarrow 1t_1$ and $3t_2 \leftarrow 1t_1$, and the energies of these states are shown in the left-hand part of Fig. 3.4. The effects of configuration interaction between all singly excited $^1A_1$, $^1T_1$, $^1T_2$ and $^1E$ states leads to the pattern of energy levels shown in the right-hand part of Fig. 3.4. The sizable effect of the configuration interaction, taken with the large number of configurations required, is probably a direct manifestation of the poor quality of the antibonding molecular orbitals (see Section 1.5). Also, in view of the discussion above, it is interesting to discover that the orbital energy differences $\epsilon_{2e} - \epsilon_{1t_1}$ and $\epsilon_{3t_2} - \epsilon_{1t_1}$ are 1·81 aJ and 2·52 aJ, respectively; results which show the irrelevance of these particular orbital energy differences for predicting transition energies. The real problem, in this instance, arises because the single electron excitation $2e \leftarrow 1t_1$, for example, leads to singlet $T_1$, $T_2$ excited states which differ in energy by a significant amount ($\hat{H}_2$ is the part of the Hamiltonian which is responsible for the term splitting).

If the metal–ligand interaction is weak, or of intermediate strength, it is not obvious whether the effects of $\hat{H}_2$ or $\hat{V}$ will be dominant. In these circumstances, it may not be possible to predict, with certainty, the spatial symmetry

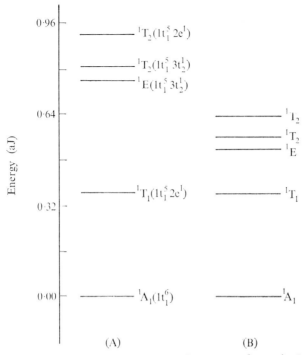

FIG. 3.4. A schematic representation of the low-energy electronic states of $MnO_4^-$ as calculated by Brown, James, O'Dwyer and Roby with (B) and without (A) configuration interaction. In both cases the zero of energy is the lowest energy $^1A_1$ state.

and spin multiplicity of the ground-state molecular term. It is necessary, therefore, to perform preliminary calculations on all terms arising from the configurations $t_{2g}^{6-m}e_g^m$, in order to discover which term, out of which configuration, has the lowest energy. A configuration interaction calculation may then be required to improve the wave functions and transition energies still further.

After these calculations have been made, it may well be found, for example, that the configuration with a maximum number of paired electrons in the $t_{2g}$ molecular orbitals does not give rise to the molecular term of lowest energy. This result follows because, if $\Delta'$ is small enough, the large intra-molecular orbital Coulomb repulsion energy can be reduced by promoting electrons into the empty $e_g$ molecular orbitals. For example, the configuration $t_{2g}^4 e_g^2$ gives rise to a $^5T_{2g}$ term which, in addition to having a reduced intra-molecular orbital repulsion energy, is also favoured by an increased exchange energy, since the spins or four previously paired electrons can now be unpaired. It is therefore not surprising to find some octahedral $d^6$ complexes characterized by a ground state of high spin multiplicity; for example, the hexafluoro anions of $Co^{+++}$ and $Fe^{++}$.

The existence of these "high spin" complexes, with ground states associated with open-shell electron configurations, present a new, and interesting, situation because, from the experimental point of view, complexes of this kind are expected to show characteristic magnetic properties, as well as a complex pattern of spectral absorption bands. In fact, it was one of the early successes of crystal field theory, as developed by Tanabe and Sugano,[22] and Kotani[23] which enabled many of the experimental magnetic and optical properties to be correlated and understood. However, as already discussed at the end of Section 3.5, the effects of covalency cannot be treated within the framework of the crystal field model. It is therefore of some interest to discover whether some form of molecular orbital theory can provide a more useful approach to the problem. This forms the subject of Section 3.7.

## 3.7.  The problem of complex ions with open-shell electronic structures

The methods of handling molecular orbital wave functions for open-shell electron configurations have been briefly alluded to in Chapter 1, and a further discussion will be found in Vol. 1.

As indicated in Chapter 1, there are many non-trivial problems associated with the calculation of open-shell wave functions. It is not surprising, therefore, to learn that very few applications of any form of open-shell theory have been made to inorganic complexes; but this situation will almost certainly be ameliorated in the very near future, with the advent of improved computational facilities.

Several attempts have been made to develop semi-empirical models, which are based on the Hückel assumption of using a single effective Hamiltonian, the form of which remains unspecified. However, as already seen in Chapter 1, any approach of this kind is gravely deficient because it avoids detailed, consideration of the two-electron terms in $\hat{H}$ (the classic application of Hückel theory to planar unsaturated hydrocarbons yields useful results only so long as there is no marked polarization of the electronic charge distribution). Hückel methods are, of course, totally inadequate for discussing excited states of heteropolar molecules; and these are the properties of most interest as far as transition metal ion complexes are concerned.

Fenske and Radtke[24] have developed a form of Hückel theory for calculating the electronic structure of $VCl_6^{---}$, $TiCl_6^{---}$, $CrCl_6^{---}$ and other complex ions. If the deficiencies of the Hückel approach are disregarded for the moment, it is interesting to discover that the two molecular orbitals, whose separation in energy is $\Delta'$, are both mainly concentrated in the region of the transition metal ion. This would indicate that an electronic transition, arising from an electron jump between these two levels, is essentially a $d$–$d$

transition. Unfortunately, the authors made no attempt to assess the importance of configuration interaction, nor was any attempt made to deal adequately
with the valence-core non-orthogonality problem.

Roos[25] has developed a semi-empirical form of the Roothaan open-shell
theory. The only drawback with his method is in the great plethora of
parameters which arise, and it is too early to judge whether the method will
be of much use. Roos has given results for the valence electronic structures of
$Cu(NH_3)_6^{++}$ and $Cu(H_2O)_6^{++}$—again omitting the effects of valence-core non-
orthogonality—and he finds, as expected, that the ground state is strongly
influenced by the configuration interaction with singly excited states; but
similar calculations were not reported for the approximate excited states.

A more recent, and comprehensive, unrestricted calculation has been made
by Ellis and co-workers[26] on the $NiF_6^{----}$ aggregate which exists in $KNiF_3$.
The separation in energy between the $^3A_{2g}(t_{2g}^6 e_g^2)$ and $^3T_{2g}(t_{2g}^5 e_g^3)$ electronic
states, which is usually ascribed to $\Delta$ in the phenomenological crystal field
approach, is found to be $0.208$ aJ, compared with the value of $0.144$ aJ which
can be deduced from the experimental results (note that the two results are not
strictly comparable, as the experimental data are appropriate to the crystalline solid). The energies and sizes of the $e_g$ and $t_{2g}$ one-electron orbitals, used
for accommodating the valence electrons, are also sensitive to the spin state of
the occupying electron. For example, in an obvious notation the orbital
energies for the $^3A_{2g}$ state are

$$\epsilon(e_g; \epsilon, \alpha) = \epsilon(e_g; \theta, \alpha) = 3.570 \text{ aJ},$$

$$\epsilon(t_{2g}; \xi, \alpha) = \epsilon(t_{2g}; \eta, \alpha) = \epsilon(t_{2g}; \zeta, \alpha) = 3.452 \text{ aJ},$$

$$\epsilon(t_{2g}; \xi, \beta) = \epsilon(t_{2g}; \eta, \beta) = \epsilon(t_{2g}; \zeta, \beta) = 3.491 \text{ aJ},$$

while those for the $^3T_{2g}$ state are

$$\epsilon(e_g; \epsilon, \alpha) = 3.583 \text{ aJ}, \quad \epsilon(e_g; \theta, \alpha) = 3.539 \text{ aJ}.$$

$$\epsilon(e_g; \epsilon, \beta) = 3.657 \text{ aJ},$$

$$\epsilon(t_{2g}; \zeta, \alpha) = 3.408 \text{ aJ},$$

$$\epsilon(t_{2g}; \eta, \alpha) = \epsilon(t_{2g}; \xi, \alpha) = 3.430 \text{ aJ},$$

$$\epsilon(t_{2g}; \xi, \beta) = \epsilon(t_{2g}; \eta, \beta) = 3.469 \text{ aJ}.$$

These results indicate that the conventional description of the electronic
structure of transition metal ion complexes, using the same spatial orbitals for
electrons of either spin, may require a radical revision—particularly when the
ground or excited electronic states are described by open-shell electron
configurations—because, in these instances, the splitting of the $e_g$ spin orbital
energies is of the same order of magnitude as $\Delta'$. In defence of the more
traditional approach, however, it must be said that the authors made no
attempt to generate a pure spin multiplet for the excited state; nor was the

effect of configuration interaction considered for the excited state (in the ground state, with a half-filled shell, the orbitals can be chosen to transform like the irreducible representations of $O_h$ without loss of generality; but this is not the case in the excited state, where the choice of symmetry adapted one-electron orbitals represents a restriction).

Several other attempts to calculate the extent of ligand orbital participation in transition metal ion complexes, or aggregates, have been made (see Bibliography), but it is too premature to say which method will turn out to be most useful. It is therefore with some reluctance that one must remain satisfied with existing phenomenological approaches for the time being, as the basic theory is still in need of considerable development. In fact it seems likely that a greater degree of success will be forthcoming from applications of valence bond theory; but such advances will have to await the development of more efficient handling of the non-orthogonality problem.

## REFERENCES

1. BETHE, H., *Ann. Phys.*, 1929, **3**, 133.
2. ZARE, R. N., *J. Chem. Phys.*, 1967, **47**, 3561.
3. GRIFFITH, J. S., *The Theory of Transition Metal Ions*, 1964, Cambridge University Press.
4. MARGENAU, H. and MURPHY, G. M., *The Mathematics of Physics and Chemistry*, 1956, Van Nostrand, New York.
5. VAN VLECK, J. H., *J. Chem. Phys.*, 1939, **7**, 72.
6. LIEHR, A. D., *Z. f. Naturforschung*, 1958, **13a**, 311.
7. McWEENY, R., *Symmetry—An Introduction to Group Theory and its Applications*, 1963 (Topic 1, volume **3** of this series), Pergamon Press, Oxford.
8. WILSON, E. B., JR., DECIUS, J. C. and CROSS, P. C., *Molecular Vibrations*, 1955, McGraw-Hill Book Co., New York.
9. LIEHR, A. D. and BALLHAUSEN, C. J., *Phys. Rev.*, 1957, **106**, 1161.
10. KOIDE, S. and PRYCE, M. H. L., *Phil. Mag.*, 1958, **3**, 607.
11. MOORE, C. E., *Atomic Energy Levels*, 1949, Circular No. 467, National Bureau of Standards, Washington, DC.
12. CONDON, E. U. and SHORTLEY, G. H., *The Theory of Atomic Spectra*, 1935, chapter 3, Cambridge University Press.
13. HARTMAN, H. and FURLANI, C., *Z. f. Phys. Chem.*, 1956, **9**, 162.
14. LOW, W., *Z. f. Phys. Chem.*, 1957, **13**, 107.
15. KLEINER, W. H., *J. Chem. Phys.*, 1952, **20**, 1784.
16. TANABE, Y. and SUGANO, S., *J. Phys. Soc. (Japan)*, 1956, **11**, 864.
17. PHILLIPS, J. C., *J. Phys. Chem. Solids*, 1959, **11**, 226.
18. JØRGENSEN, C. K., *Absorption Spectra and Chemical Bonding in Complexes*, 1962, chapter 7, Pergamon Press, Oxford.
19. SHULMAN, R. G. and SUGANO, S., *Phys. Rev.*, 1963, **130**, 506.
20. PEYERIMHOFF, S. D., 1967, *J. Chem. Phys.*, 1967, **47**, 349.
21. BROWN, R. D., JAMES, B. H., O'DWYER, M. F., and ROBY, K. R., *Chem. Phys. Letters*, 1967, **1**, 459.
22. TANABE, Y. and SUGANO, S., *J. Phys. Soc. (Japan)*, 1954, **9**, 753, 766.
23. KOTANI, M., *J. Phys. Soc. (Japan)*, 1949, **4**, 293.
24. FENSKE, R. F. and RADTKE, D. D., *Inorg. Chem.*, 1968, **7**, 479.
25. ROOS, B., *Acta Chem. Scand.*, 1966, **20**, 1673.
26. ELLIS, D. E., FREEMAN, A. J. and ROS, P., *Phys. Rev.*, 1968, **176**, 688.

# BIBLIOGRAPHY

FRIEDMAN, H. G., Jr., CHOPPIN, G. R. and FEUERBACHER, D. G., On the shapes of $f$ orbitals, *J. Chem. Ed.*, 1964, **41**, 354.

HUTCHINGS, M. T., Point charge calculations of energy levels of magnetic ions in crystalline electric fields, *Solid State Physics*, 1964, **16**, 227.

LIEHR, A. D., The coupling of vibrational and electronic motions in degenerate electronic states of inorganic complexes, *Adv. Inorg. Chem.*, 1962, **3**, 281; 1962, **4**, 455; 1963, **5**, 385.

LIEHR, A. D. and BALLHAUSEN, C. J., Inherent configurational instability of octahedral inorganic complexes in $E_g$ electronic states, *Ann. Phys.*, 1958, **3**, 304.

MOFFITT, W. E. and THORSON, W., Vibronic states of octahedral complexes, *Phys. Rev.*, 1957, **108**, 1251.

YERANOS, W. A., On the theory of vibronic interactions, *Zeits. f. Naturforschung*, 1967, **22a**, 183.

LIEHR, A. D. and BALLHAUSEN, C. J., Complete theory of Ni(II) and V(III) in cubic crystalline fields, *Ann. Phys.*, 1959, **2**, 134.

HERZFELD, C. M. and GOLDBERG, H., On the nature of the crystal field approximation, *J. Chem. Phys.*, 1961, **34**, 643.

JARRETT, H. S., Generalization of crystal field theory to include covalent bonding, *J. Chem. Phys.*, 1959, **31**, 1579.

WATSON, R. E. and FREEMAN, A. J., Covalent effects in rare-earth crystal field splittings, *Phys. Rev.*, 1967, **156**, 251.

ELLIS, M. M. and NEWMAN, D. J., Crystal field in rare-earth trichlorides. I. Overlap and exchange effects in $PrCl_3$, *J. Chem. Phys.*, 1967, **47**, 1986.

BEDON, H. D., HORNER, S. M. and TYREE, S. Y., Jr., A molecular orbital treatment of the spectrum of the hexafluorotitanium III anion, *Inorg. Chem.*, 1964, **3**, 647.

DAHL, J. P., On the application of ZDO methods to complex molecules, *Acta Chem. Scand.*, 1967, **21**, 1244.

DAHL, J. P. and BALLHAUSEN, C. J., Molecular orbital theories of inorganic complexes, *Adv. Quantum Chem.*, 1968, **4**, 170.

BASCH, H., VISTE, A. and GRAY, H. B., Molecular orbital theory for octahedral and tetrahedral metal complexes, *J. Chem. Phys.*, 1966, **44**, 10.

SUGANO, S. and TANABE, Y., Covalency in ionic crystals: $KNiF_3$, *J. Phys. Soc. (Japan)*, 1965, **20**, 1155.

HUBBARD, J., RIMMER, D. E. and HOPGOOD, F. R. A., Weak covalency in transition metal salts, *Proc. Phys. Soc.*, 1966, **88**, 13.

BALLHAUSEN, C. J., *Introduction to Ligand Field Theory*, 1962, McGraw-Hill Book Co. Inc., New York.

JØRGENSEN, C. K., Recent progress in ligand field theory, *Structure and Bonding*, 1966, **1**, 3.

OWEN, J. and THORNLEY, J. H. M., Covalent bonding and magnetic properties of transition metal ions, *Rep. Prog. Phys.*, 1966, **29**, 675.

ANDERSON, P. W., Theory of exchange in insulators (Section 6), *Solid State Physics*, 1963, **14**, 99.

KOIDE, S. and GONDAIRA, K. I., Recent developments in the theory of transition metal ions, *Prog. Theor. Phys. Supplement*, 1967, **40**, 160.

McCLURE, D. S., Electronic spectra of molecules and ions in crystals, *Solid State Reprint*, 1959, Academic Press.

DUNN, T. M., McCLURE, D. S. and PEARSON, R. G., *Some Aspects of Crystal Field Theory*, 1965, Harper & Row, New York.

FIGGIS, B. N., *Introduction to Ligand Fields*, 1966, Interscience, New York.

BERTHIER, G., Self-consistent field methods for open-shell molecules, *Molecular Orbitals in Chemistry, Physics and Biology*, 1964, Academic Press, New York.

McWEENY, R., The density matrix in self-consistent field theory. III. Generalizations of the theory, *Proc. Roy. Soc.*, 1967, A **241**, 239.

CHAPTER 4

# THE ELECTRONIC STRUCTURE OF XENON FLUORIDES

## 4.1. Introduction

The isolation of molecules containing a rare gas atom poses fascinating chemical and quantum mechanical problems, as the existence of these molecules completely contravenes previously established ideas associated with the chemical inactivity of the rare gases. However, it is interesting that Pimentel[1] anticipated the existence of $XeF_2$ and $XeF_4$ in 1951, eleven years before they were prepared, by noticing that they were isoelectronic (at least as far as the valence electrons are concerned) with the polyhalide ions $ICl_2^-$ and $ICl_4^-$, respectively.

$XeF_4$ was in fact the first molecule containing a rare gas atom to be prepared,[2] and it was found to have the square planar structure. $XeF_2$ is linear; and $XeF_6$ is thought to be nearly octahedral, but the experimental data have not yet been completely rationalized. For convenience, some of the existing experimental data, pertaining to the ground state geometries of the di-, tetra- and hexa-fluorides of xenon, are summarized in Table 4.1.

TABLE 4.1

*A summary of the experimental geometries of the di-, tetra- and hexa-fluorides of xenon*

| Molecule | Shape | $R_{XeF}(nm)$ | Method | Refs. |
|----------|-------|---------------|--------|-------|
| $XeF_2$ | linear | $0 \cdot 214 \pm 0 \cdot 014$ | X-ray | 3 |
| | | $0 \cdot 200 \pm 0 \cdot 001$ | neutron diffraction | 4 |
| | | $0 \cdot 1977 \pm 0 \cdot 00015$ | spectroscopic | 5 |
| $XeF_4$ | square planar | $0 \cdot 1965 \pm 0 \cdot 0022$ | X-ray | 6 |
| | | $0 \cdot 195 \pm 0 \cdot 001$ | neutron diffraction | 7 |
| | | $0 \cdot 194 \pm 0 \cdot 001$ | electron diffraction | 8 |
| $XeF_6$ | nearly octahedral (?) | $0 \cdot 1890 \pm 0 \cdot 0005\dagger$ | electron diffraction | 9 |

† This is the mean Xe–F bond distance; the experimental data are not compatible with a regular octahedral molecule vibrating in independent normal modes.[9]

In order to obtain an understanding of the bonding in these molecules, it is necessary to discover why fluorine is so successful in stabilizing complexes containing xenon; and, secondly, why it is that only certain stereochemical arrangements of fluorine ligands are observed. Unfortunately, at the present time, it is possible to provide only qualitative answers to these questions, in view of the dearth of suitable quantitative studies on the electronic structure of the rare gas fluorides. The remainder of this chapter is therefore concerned mainly with a qualitative discussion of the bonding, in terms of the simple valence bond and molecular orbitals models introduced in Chapter 2. The valence bond approach is considered first.

## 4.2.  The valence bond model

(a) *Xenon difluoride*, $XeF_2$. There are five pairs of valence electrons to be considered in xenon difluoride, if the electrons in the $\pi$-orbitals on the fluorine ligands are assumed to take no part in the bonding. The conventional Pauling–Slater approach to the bonding would therefore require the construction of $sp^2pd$ trigonal bipyramidal hybrids; with the two $pd$ hybrids used for bonding the two fluorine atoms in the linear configuration, as shown in Fig. 4.1. The $\sigma$-hybrid orbitals used by fluorine are predominantly $p$-like in character, since the $p \leftarrow s$ promotion energy is large.

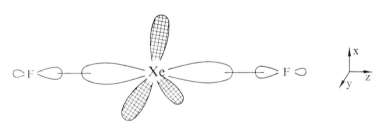

Fig. 4.1. A chemically motivated singlet-coupled covalent valence bond structure for $XeF_2$, based on the use of $p_z d_{z^2}$ digonal hybrid atomic orbitals by xenon. Interatomic electron pairing is indicated by a line joining singly occupied orbitals, and the three hybrid orbitals, $tr_i$ ($i = 1, 2, 3$), containing lone-pairs of electrons are indicated by crossed hatching.

Now the use of hybrid orbitals by xenon requires careful justification, in so far as some estimate must be made for the promotion energy involved. In fact, this promotion energy is associated only with the occupancy of the $pd$ hybrids, as no promotion energy is required for doubly occupying the three xenon $sp^2$ hybrids having their maximum amplitudes in the $xy$-plane—a result which is best seen by expanding the lone-pair part of the valence bond wave function in terms of ordinary atomic orbitals. Thus, with the axis system as defined in Fig. 4.1, the three hybrid atomic orbitals are given by

$$tr_1 = \frac{1}{\sqrt{3}}(s + \sqrt{2}.p_x); \quad tr_2 = \frac{1}{\sqrt{3}}\left(s - \frac{1}{\sqrt{2}}p_x + \sqrt{\frac{3}{2}}.p_y\right);$$

$$tr_3 = \frac{1}{\sqrt{3}}\left(s - \frac{1}{\sqrt{2}}p_x - \sqrt{\frac{3}{2}}.p_y\right).$$

And the approximate ground state wave function, $\Psi_0$, originally given by (only the lone-pair part is shown explicitly)

$$\hat{A}tr_1(1)\overline{tr_1}(2)tr_2(3)\overline{tr_2}(4)tr_3(5)\overline{tr_3}(6) \ldots$$

therefore becomes

$$\left(\frac{1}{\sqrt{3}}\right)^6 \hat{A}(s + \sqrt{2}.p_x)(1)\left(s - \frac{1}{\sqrt{2}}p_x + \sqrt{\frac{3}{2}}.p_y\right)(2) \times$$

$$\times \left(s - \frac{1}{\sqrt{2}}p_x - \sqrt{\frac{3}{2}}.p_y\right)(3)(\bar{s} + \sqrt{2}.\overline{p_x})(4) \times$$

$$\times \left(\bar{s} - \frac{1}{\sqrt{2}}\overline{p_x} + \sqrt{\frac{3}{2}}\overline{p_y}\right)(5)\left(\bar{s} - \frac{1}{\sqrt{2}}\overline{p_x} - \sqrt{\frac{3}{2}}.\overline{p_y}\right)(6) \ldots$$

$$= \left(\frac{1}{\sqrt{3}}\right)^6 \hat{A}(s + \sqrt{2}.p_x)(1)(\sqrt{3}.p_x(2)p_y(3) + \sqrt{6}.p_y(2)s(3) \times$$

$$\times (\bar{s} + \sqrt{2}.\overline{p_x})(4)(\sqrt{3}.\overline{p_x}(5)\overline{p_y}(6) + \sqrt{6}.\overline{p_y}(5)\bar{s}(6)) \ldots$$

$$= \hat{A}s(1)\bar{s}(2)p_x(3)\overline{p_x}(4)p_y(5)\bar{p}_y(6) \ldots .$$

This result also shows that the trigonal disposition of the lone-pairs of electrons, as indicated in Fig. 4.1, has no physical significance, as the corresponding wave function is identical to the one involving the double occupancy of the $5s$ and $5p_x$, $5p_y$ atomic orbitals of xenon.

However, the involvement of xenon $pd$ hybrid atomic orbitals, for describing the bonding to the fluorine atoms, is a different matter as the necessary promotion energy is expected to be large. For instance, the excitation energy for the process $5s^25p^55d^1 \leftarrow 5s^25p^6$ is about 1·6 aJ and, since the promotion energy will contain contributions from excitation energies for states of xenon containing one or two $5d$ electrons (see the analysis for the use of $sp$ hybrids in Section 1.4), the promotion energy is expected to be somewhat greater than 1·6 aJ. The precise value of the energy clearly depends upon the extent of $p$–$d$ mixing. But, in any case, the expansion of the structure wave function in terms of ordinary atomic orbitals, will show a preponderance of structures based on the excited . . . $5d^1$ electron configuration of xenon.

Now the results of some Hartree–Fock calculations, on the $5p^55d^1$ valence electron configuration of xenon[10] (only the average term energy was optimized), show that the $5d$ orbital is reasonably compact, with the main maximum in the radial distribution function occurring at about 0·22 nm: a result which should be contrasted with the situation in sulphur, where the ... $3d^1$ configuration is associated with a much more diffuse $d$-atomic orbital (see Table 2.1). The effects of orbital contraction are therefore still expected to be important; but perhaps to a lesser degree than in sulphur, providing that the Hartree–Fock orbitals are used in the molecular calculation. Unfortunately, no calculations of this kind have yet been performed, but there are some interesting results produced by Mitchell[11] in model calculations on xenon and xenon difluoride. He finds that the xenon $5d_{z^2}$ atomic orbital, represented by a single Slater orbital suitably orthogonalized to the remaining atomic orbitals in the molecule, is strongly contracted in the field of the two fluorine atoms, thereby making it suitable for the formation of $pd$ hybrids. Mitchell also finds a value of about 1·4 aJ for the promotion energy of xenon in the trigonal bipyramidal configuration, and estimates that this energy will be regained on the formation of two xenon–fluorine bonds.

Although the single valence bond structure depicted in Fig. 4.1 is nominally described as a covalent structure, the use of xenon $pd$ hybrids enables charge to be removed from the vicinity of the central atom; this follows because the centroids of the hybrid atomic orbitals are displaced towards the ligands, in contrast to ordinary atomic orbitals whose centroids coincide with the xenon nucleus. The use of hybrid orbitals therefore helps in describing the polarization, or change in shape, of the xenon atom which is induced by the field of the fluorine ligands. The extent of the polarization can be determined only by actual calculation and, in fact, it may also be necessary to include participation by other high-energy atomic orbitals; for example, $4f$ or $6s$ atomic orbitals.

An alternative description of the bonding can be given which does not require the use of $d$, or higher energy, atomic orbitals in the first instance; this involves an application of the model already introduced in Chapter 2, for discussing the stereochemistry of molecules containing a central second row atom.

Since $XeF_2$ contains ten valence electrons, it is anticipated that the structure will be based on the trigonal bipyramid—for this was the stereochemistry successfully utilized in the discussion of the bonding in $ClF_3$ and $SF_4$ which also contain ten valence electrons (for $XeF_2$, nothing new is gained by considering the square pyramidal structure). Thus, the three $sp^2$ hybrids on xenon are occupied by three lone pairs, and the bonding occurs through the remaining $p$ atomic orbital, directed along the molecular axis. But as there are two electrons in this orbital in the free atom, one electron must be transferred to one of the fluorine atoms, so that an electron-pair bond can be formed with the other fluorine atom, as shown in Fig. 4.2(A). The alternative structure, in which an electron is transferred from xenon to the right-hand fluorine atom, is shown in Fig. 4.2(B). The approximate ground state wave function is there-

(A)                                                    (B)

Fig. 4.2. Valence bond structures for $XeF_2$ which exclude participation by xenon $5d$ atomic orbitals. The two structures are based on the use of the xenon $5p_\sigma$ and fluorine $\sigma$-hybrid (predominantly $p$-like in character) atomic orbitals. The lone-pairs of electrons in the xenon $5s$ and $5p_\pi$ and the fluorine $2p_\pi$ atomic orbitals are not shown. Interatomic electron pairing is indicated by a line joining singly occupied orbitals, and orbitals containing a lone-pair of electrons are indicated by crossed hatching.

fore described by the (normalized) symmetric combination of the two wave functions corresponding to the two structures in Fig. 4.2—and it should be remembered that these structure wave functions are not orthogonal.

The $s$, $p$ orbital-only model, in common with the hybrid model, also predicts that charge will be transferred from the central xenon atom to the ligands—but now the mechanism of charge displacement is perhaps described in more familiar chemical terms.

There is one structure, involving the four orbitals under consideration, which does not require a formal transfer of charge; this is the structure shown in Fig. 4.3. But since this is a formally long-bonded structure, it is expected to contribute only a small weight to the total molecular wave function.

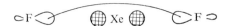

Fig. 4.3. Another singlet paired valence bond structure for $XeF_2$, based on the use of the same atomic orbitals as in Fig. 4.2. Interatomic electron pairing is indicated by a line joining singly occupied orbitals, and the orbital containing the lone-pair of electrons is indicated by crossed hatching.

The only other xenon–fluorine configurations which might be thought feasible are those involving one axial and one equatorial fluorine atom; these would give rise to an F–Xe–F bond angle of 90°. However, it is very easy to see that structures of this kind are energetically unfavourable, as they all require a doubly occupied xenon atomic orbital to penetrate the charge distribution associated with a F⁻ species.

In parallel with the discussion of molecules containing a central second-row atom, the ground-state wave function can be improved, if desired, by allowing for the presence of structures based on the use of $d$-orbitals—but now the weights of these structures are determined variationally; and they need only be included to improve the shape of the electron distribution around the xenon atom, without altering the basic stereochemistry. For example, structures involving the $5d_{z^2}$ atomic orbital on xenon can be discussed in terms of structures based on $pd$ hybrids, providing all spin couplings, and orbital occupations, are considered (see Section 1.4). In the present instance, two obvious covalent structures, based on singly occupied orbitals, are represented in Fig. 4.4; and one of these happens to be the same structure that appeared in the Pauling hybrid approach discussed above.

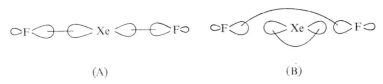

(A)                                         (B)

FIG. 4.4. Two singlet coupled valence bond structures for $XeF_2$, based on the use of $pd$ hybrid atomic orbitals by xenon. Interatomic pairing is indicated by a line joining singly occupied orbitals.

The extended valence bond calculation, involving all the structures displayed in Figs 4.2, 4.3 and 4.4, as well as any ionic structures which are also considered to be important, is not very easy to perform, because of the problems associated with the atomic orbital non-orthogonality (the structures themselves are also non-orthogonal). However, a few preliminary results are available from the previously cited work of Mitchell.[11] He has computed the diagonal matrix elements of the total Hamiltonian for the structures shown in Figs. 4.2(A) and 4.4(A). The $pd$ hybrid structure is found to have the lower energy; but it is difficult to draw any definite conclusion from this result until the effect of configuration interaction between the degenerate structures in Fig. 4.2 has been included—and this effect is unlikely to be small. Nevertheless, it is interesting that the results of these model calculations suggest that $5d$ orbital participation is at least feasible.

(b) *Xenon tetrafluoride*, $XeF_4$. Both the hybrid and the $s$, $p$ orbital only model can be used for discussing the xenon–fluorine bonding in $XeF_4$.

In the hybrid model, depicted in Fig. 4.5, $sp^2d$ square planar hybrids are used by xenon for bonding with the four fluorine atoms. The two lone-pairs are then accommodated in the $p_z d_{z^2}$ hybrids, perpendicular to the molecular plane, so as to minimize the various electron-pair repulsions.[12-13] However, as the $s$ and $d_{z^2}$ atomic orbitals both transform according to the $a_{1g}$ irreducible representation of $D_{4h}$, it is not possible to separate their relative contributions to the bonding; and it is only for simplicity that the whole of the $s$ character is assigned to the hybrids in the $xy$-plane.

The use of xenon hybrid orbitals leads again to a spreading out of the electron density towards the ligands. But, unfortunately, no calculations are

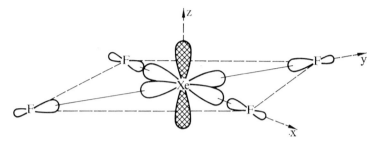

Fɪɢ. 4.5. A valence bond structure for XeF$_4$ based on the use of $sp^2d$ (square planar) and $p_zd_{z^2}$ (digonal) hybrid atomic orbitals of xenon. Interatomic electron pairing is indicated by a line joining singly occupied orbitals, and orbitals occupied by lone-pairs are indicated by crossed hatching.

currently available to give estimates of the relative sizes of the $5d$ orbitals in this molecule; so it is difficult to assess the reality of the polarization effect. Also, there is still the uncertainty as to whether the promotion energy for xenon can be regained on bond formation—a point which can be confirmed only by actual calculation, and this is one of the weaknesses of the hybrid model. However, in view of Mitchell's preliminary calculations on XeF$_2$, it does not seem unreasonable to anticipate that a similar situation will obtain for XeF$_4$.

The $s$, $p$ orbital only model, for describing the xenon-fluorine bonding, now requires two electrons to be removed from xenon, in order that two bonds can be formed with fluorine atoms as shown in Fig. 4.6. This structure, in

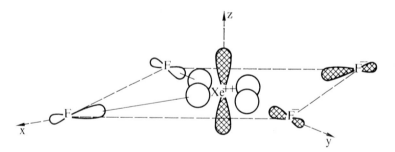

Fɪɢ. 4.6. A possible valence bond structure for XeF$_4$ based on Xe$^{++}$. The $p_x$, $p_y$ atomic orbitals of xenon are used for bonding, and the $sp_z$ digonal hybrids accommodate two lone-pairs of electrons. Interatomic electron pairing is indicated by a line joining singly occupied orbitals, and orbitals occupied by lone-pairs are indicated by crossed hatching.

which the $sp_z$ hybrids are used to accommodate two lone-pairs of electrons, exactly parallels the one used in discussing the electronic structure of SF$_6$ (a twelve-valence electron molecule). And, once again, the energy required for the transfer of the two electrons is expected to be mostly offset by the net Coulomb attraction between the separated formal charges.

There are four possible structures of the kind shown in Fig. 4.6, and the ground state is expected to be represented by the linear combination of structures transforming like the totally symmetric irreducible representation of $D_{4h}$. The remaining three linear combinations of structures, transforming like other (non-totally symmetric) irreducible representations of $D_{4h}$, are expected to lie higher in energy. It must be remembered, of course, that other high-energy structures can be formulated using different spin pairing schemes, but these are ignored in the present discussion.

It is now easy to see that the tetrahedral configuration of fluorine ligands, based on structures like the one shown in Fig. 4.7, is energetically unfavourable, because the lone-pairs of electrons on xenon are forced to occupy regions

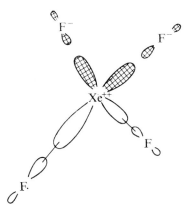

FIG. 4.7. A valence bond structure for XeF$_4$ based on the tetrahedral configuration of fluorine ligands, and the use by xenon of $sp^3$ hybrid atomic orbitals. Interatomic electron pairing is indicated by a line joining singly occupied orbitals, and orbitals occupied by lone-pairs are indicated by crossed hatching.

of space which are already occupied by the fluorine lone-pair electrons. The molecule would therefore distort, presumably into the planar configuration, so as to relieve the excess energy of repulsion.

Structures involving the occupancy of $d$-orbitals can now be introduced to improve the basic description of the electronic charge distribution, in exactly the same way as in the earlier discussion of SF$_6$. It will be remembered that the contribution of structures of this kind tend to reduce the extent of formal charge transfer between the central atom and the ligands. For example, one possible structure is obtained by transferring one of the $\sigma$-lone-pair electrons of a F$^-$ ion (see Fig. 4.6) into the $d_{x^2-y^2}$ atomic orbital of xenon. However, this structure by itself is not symmetry adapted, and the appropriate linear combination of equivalent structures, transforming like the totally symmetric irreducible representation of $D_{4h}$, is found by the standard group theoretical techniques already discussed in some detail. Now each structure, in this linear combination of structures, is characterized by a formal charge of $+1$ on the central xenon atom; this is fortunate as it also ensures that the $d$-orbital will be of a suitable size, and energy, for enabling some charge to be back-donated from the fluorine ligands to the xenon atom. But it should be remembered, of course, as already discussed in Chapters 1 and 2, that the apparent transfer of charge is only formal, because the change in electron density distribution will depend upon the size of the $d$-orbitals. Nevertheless, it is still useful to regard these structures as contributing towards the reduction in the extent of formal charge transfer.

The structure just described is clearly not the only one that can be formulated in terms of singly (or doubly) occupied $d$-orbitals; but it is thought to contribute most towards the improvement of the ground state wave function.

This completes the discussion of the valence bond description of the bonding in $XeF_4$, and leads finally to the problem of $XeF_6$.

(c) *Xenon hexafluoride*, $XeF_6$. As mentioned in the introduction, the molecular structure of $XeF_6$ is still not known with certainty, although it

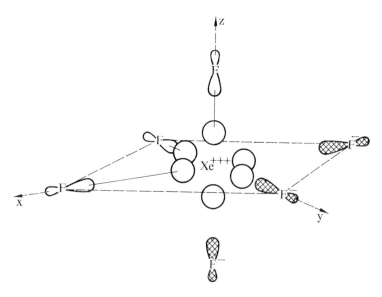

FIG. 4.8. A typical valence bond structure for $XeF_6$ based on $Xe^{+++}$. The orbitals used by xenon for bonding are $p_x$, $p_y$ and $p_z$: the $5s$ atomic orbital accommodates a lone-pair of electrons, and is not shown. Interatomic electron pairing is indicated by a line joining singly occupied orbitals, and orbitals containing lone-pairs are indicated by crossed hatching.

appears to be nearly octahedral. For the purposes of the present discussion, therefore, it is useful to assume initially that the molecule has a regular, octahedral structure. In this situation, the simple $s$, $p$ orbital-only model requires a formal displacement of three electrons from xenon to the ligands, in order to accommodate the necessary number of valence electrons. For example one possible structure, shown in Fig. 4.8, involves the use of the xenon $5s$ atomic orbital to accommodate a lone-pair of electrons, and the three singly occupied pure $5p$ atomic orbitals are then available for forming electron-pair bonds with three fluorine atoms. However, alternative structures can be formulated, if desired, by using $sp$ hybrid orbitals on xenon; but now all orbital occupancies, and spin couplings, must be considered in order to relax the constraints inherent in the use of hybrid orbitals. But, in view of the large $p \leftarrow s$ promotion energy in xenon, it seems more reasonable, perhaps, to ignore any xenon $s$–$p$ atomic orbital mixing.

Now any structure of $XeF_6$, based on the use of xenon $5s$ and $5p$ atomic orbitals, will involve an unnecessarily large amount of formal charge transfer; and this must be considered as a possible limitation of the simple model. However, as will become evident shortly, it is the existence of such a limitation which is indicative of a more interesting, and complex, electronic structure for $XeF_6$.

$XeF_6$ differs from both $XeF_2$ and $XeF_4$ in that the $5p$ atomic orbitals ($t_{1u}$ orbitals in $O_h$ symmetry) on xenon are singly occupied in the simple structure shown in Fig. 4.8—compare with Figs. 4.1 and 4.6. Thus, within the framework of the simple valence bond model, as advocated here and in Chapter 2, it is now energetically feasible to reduce the large amount of formal charge transfer by considering structures based on the occupancy of either $\sigma - (e_g)$ or $\pi - (t_{2g})$ $d$-orbitals of xenon. This is in direct contrast to the situation in $XeF_2$ and $XeF_4$, where it is only worth while to consider back-donation of charge through the $\sigma$-framework; structures involving the occupancy of $d_{\pi}$-orbitals in these molecules will be energetically unfavourable because of the high energy of repulsion arising from the filled xenon $5p_{\pi}$ atomic orbitals.

It is therefore expected that structures involving participation by xenon $5d$ orbitals will dominate the wave function, as it should be easier to obtain the appropriate promotion energy for $Xe^{++}$ than to remove a further electron and exclude participation by $d$-orbitals—this should be contrasted with the situation of sulphur in $SF_6$.

Structures involving back-donation through $t_{2g}(\pi)$ orbitals are easier to envisage (see Fig. 4.9) than the corresponding structures involving xenon $e_g$ orbitals; but the former structures may not be so important, as the greater overlap between xenon and ligand $\sigma$-orbitals allows for a greater degree of back-donation. In either case, the transfer of an electron from one of the $F^-$ ions is useful in that it reduces the energy of repulsion arising from the penetration of the $F^-$ species into the xenon electron distribution.

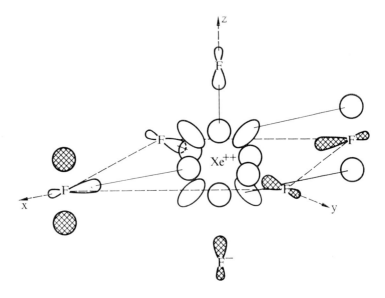

Fig. 4.9. A typical valence bond structure for XeF$_6$ based on Xe$^{++}$. The orbitals used by xenon for bonding are the $p_x$, $p_y$, $p_z$ and $d_{xz}$ atomic orbitals: the xenon $5s$ atomic orbital accommodates a lone pair of electrons (not shown). Interatomic electron pairing is indicated by a line joining singly occupied orbitals, and orbitals containing lone-pairs of electrons are indicated by crossed hatching. Only those fluorine $\pi$-atomic orbitals are shown which are involved with the back-bonding through the xenon $5d_{xz}$ atomic orbital.

Several different kinds of structure can be formulated which involve either a single or double occupancy of a xenon $t_{2g}(5d)$ atomic orbital; but those with two electrons transferred to xenon are expected to occur with small weight in the expansion of the molecular wave function. In any event, the set of structures will form a basis for a reducible representation of $O_h$, and the application of the standard group theoretical techniques will yield linear combinations of structures transforming like the irreducible representations of $O_h$.

Structures arising through the participation of xenon $e_g(5d\text{-})$ atomic orbitals are best described in terms of $pd$ hybrids. However, as already indicated previously, this scheme will allow for contributions from xenon electron configurations possessing zero, single, or double-$d$-orbital occupancy. It is therefore necessary to include the remaining non-chemical structures involving other hybrid orbital occupancies, and all singlet-spin couplings, in order to overcome the constraints inherent in the use of hybrid atomic orbitals in a valence bond approach (the same procedure ought really to be carried out to allow for the effects of $5s$–$5p$ mixing, and it is assumed that this would be accomplished in an actual calculation). A typical structure is shown in Fig. 4.10, but other equivalent structures can also be formulated which involve the $5d_{x^2-y^2}$ orbital on xenon.

The extent of formal charge transfer can be reduced still further, by allowing for an additional back-transfer from one of the two $F^-$ ions into the appropriate xenon $t_{2g}$ orbital. However, these structures now involve $Xe^+$ with an increased promotion energy, because of the increased $d$-orbital occupancy. And actual calculation can only decide on the effectiveness of these structures in improving the description of the lowest energy state. But, once again, states of several symmetries can be constructed out of the set of structures, and it requires a calculation to determine the nature of the lowest energy state; in particular, the effects of configuration interaction are expected to be

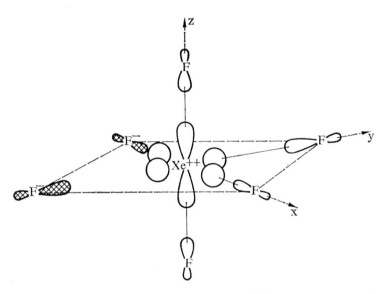

Fig. 4.10. A typical valence bond structure for $XeF_6$, based on the use of xenon $p_x$, $p_y$, and $p_z d_{z^2}$ hybrid atomic orbitals. Interatomic electron pairing is indicated by a line joining singly occupied orbitals, and orbitals containing lone pairs of electrons are indicated by crossed hatching. The xenon $5s$ atomic orbital accommodates a lone-pair of electrons, and is not shown.

important, and it is by no means certain that a spatially non-degenerate state will lie lowest in energy. The result of this calculation will almost certainly show a group of molecular states, based on the $5d^1$ and $5d^2$ electron configurations of xenon, with energies lower than the states arising from the $s$, $p$ orbital-only structures; but, of course, there is no unique identification of structure type because of the effects of configuration interaction. The presence of several low-lying molecular states, some of which will be spatially degener-

F

ate, is clearly going to have a marked influence on the electronic structure of $XeF_6$ — for even if the lowest energy state is non-degenerate, the presence of low-lying degenerate states will cause a breakdown in the assumptions implied by the use of molecular wave functions of the form (1.1) (see also (1.9)). There are therefore strong indications that a proper treatment of the electronic structure of $XeF_6$ must include the effects of vibronic coupling, caused by the presence of low-lying degenerate electronic states. This view-point has been also expressed by Bartell,[14] and by Bartell and Gavin[9] in discussions on the interpretation of the results of electron diffraction experiments on $XeF_6$.

The simple Pauling hybrid model, which requires no formal charge transfer, involves the use of $sp^3d^2$ hybrids by xenon, and the two extra electrons are then presumably accommodated in the xenon $t_{2g}$ atomic orbitals. However, the use of this single, chemically suggestive, spin-coupled wave function is likely to involve an excessive amount of promotion energy; for, on expanding the structure wave function ($2^6$ determinants) in terms of ordinary atomic orbitals, each antisymmetrized product of atomic orbitals is found to contain a minimum of two occupied xenon $5d$ atomic orbitals. And structures based on these configurations of xenon are unlikely to contribute a significant weight to the ground-state wave function.

Thus, it is easy to see that the main disadvantage of the traditional hybrid approach arises from the restriction that the ground-state wave function must be represented by a covalent structure; that is, there must be as many electron-pair bonds as there are ligands. This is why the seventh (lone) pair of electrons in $XeF_6$ must be accommodated in the xenon $t_{2g}(5d)$ atomic orbitals, as all the other valence orbitals are involved in the formation of $\sigma$-bonds to the fluorine ligands. However, as already seen above, it is much more sensible to break one or two bonds so that the extra electrons can be accommodated on the ligands. The $e_g$ orbitals are thereby allowed to spread over the molecule, rather than being localized on the xenon atom. But the ground state is now represented by a linear combination of formally ionic structures, and the model exactly parallels the one used in Chapter 2 for discussing compounds containing a central second-row atom.

Although the $sp^3d^2$ hybrid model looks energetically unfavourable, it at least predicts a non-octahedral ground state geometry for $XeF_6$. This follows because the coupling of the two electrons in the $t_{2g}$ xenon orbitals determines the symmetry of the ground state term, as the six $\sigma$-bonding pairs of electrons are necessarily coupled to yield a totally symmetric singlet state (compare with $SF_6$). But, as already seen in Chapter 3, the configuration $t_{2g}^2$ gives rise to the electronic states

$$^3T_{1g}, \ ^1A_{1g}, \qquad ^1T_{2g}, \ ^1E_g.$$

And the lowest energy state is expected to be one of the states based on the single occupancy of two of the $t_{2g}$ orbitals, so that the intra-atomic Coulomb repulsion energy is as small as possible. This would therefore favour either $^3T_{1g}$ or $^1T_{2g}$, both of which are spatially degenerate (see, for example, table A24 in Griffith). The $^3T_{1g}$ state is expected to lie lowest in energy and, even though the effects of spin-orbit coupling will presumably alleviate the spin degeneracy, the continued presence of spatial degeneracy will induce Jahn–Teller distortions.

This completes the discussion of the valence-bond model, but, before proceeding to the molecular orbital model, it is worth while digressing a little to discuss the expected stereochemistry of $XeF_5^+$, which has been characterized recently[15] in the ionic solid $(XeF_5)^+(PtF_6)^-$. The existence of this species is readily understood on the basis of the structure shown in Fig. 4.9, since the removal of a $F^-$ ion will lead to a decrease in the repulsive energy of penetration. The stereochemistry is therefore expected to be square pyramidal, in broad agreement with the experimental result.[15] It should be remembered, of course, that in the simple description given here, the $5s$ xenon atomic orbital plays no active part in the bonding. However, in the absence of one of the fluorine ligands, it seems probable that some $5s$–$5p_z$ atomic orbital mixing will be energetically worthwhile so that the lone pair can spread into the region of space previously occupied by the sixth fluorine ligand.

## 4.3. The molecular orbital model

The use of molecular orbital theory, for describing the electronic structure of the xenon fluorides, yields broadly similar conclusions to those already found from the application of valence-bond theory; but it will be seen that the molecular orbital theory provides a different, yet complementary, view of the same problems. As in the valence-bond approach, participation by xenon $5d$ atomic orbitals is initially precluded from the discussion.

(a) *Xenon difluoride*, $XeF_2$. In $XeF_2$ the atomic orbitals involved in the LCAO expansion for the valence molecular orbitals are taken as the xenon $5s$, $5p$ orbitals and suitable fluorine $\sigma$ hybrid orbitals, directed towards xenon. The fluorine $\pi$-orbitals are assumed to accommodate the four lone-pairs of electrons which play no active role in the bonding. The pattern of molecular orbital energy levels is therefore expected to be of the form shown in Fig. 4.11, which suggests a ground state electron configuration of $1\sigma_g^2 1\sigma_u^2 \pi_u^4 2\sigma_g^2$ for the lowest energy $^1A_{1g}$ state.

In this scheme the four electrons in the xenon $\pi$-orbitals make no contribution to the bonding, and if the electrons in the remaining molecular orbitals are assumed to be shared equally between xenon and the two fluorine ligands, then the effective charge on xenon is $+1$; a result which parallels the

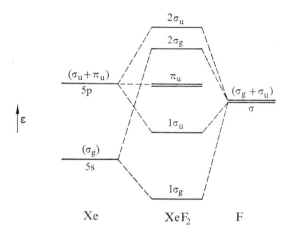

Fig. 4.11. A schematic molecular orbital energy level diagram for $XeF_2$.
$\pi$-bonding with the ligands is neglected, and xenon $5d$ atomic orbitals are
excluded from the set of expansion functions.

requirement of a formally positively charged xenon atom in the simple
valence-bond model.

The inclusion of xenon $5d(\sigma\text{-})$ atomic orbitals allows charge to be formally
back-donated to xenon through the $\sigma$-framework; the possibility of a back
transfer of charge through the $\pi$-system of atomic orbitals seems less favourable, from an energetic viewpoint, because of the large repulsive energy that
would arise between an electron in a xenon $5d(\pi\text{-})$ orbital and the filled xenon
$5p(\pi\text{-})$ shell. The effect of including the xenon $5d(\sigma\text{-})$ atomic orbital in the
LCAO expansion will therefore result in a depression of the $2\sigma_g$ molecular
orbital energy level (and, of course, the $1\sigma_g$ level, but to a much smaller
extent), and the total energy is presumably lowered in the process.

(b) *Xenon tetrafluoride*, $XeF_4$. In $XeF_4$ the pattern of energy levels must be
estimated for both the tetrahedral and square planar configurations of fluorine
ligands. Simple intuitive arguments, of the kind already used in Chapter 2,
lead to the two sequences of energy levels as shown in Fig. 4.12—and here
again, the $\pi$-orbitals on fluorine are assumed to accommodate pairs of
electrons which do not contribute to the bonding. However, the latter
assumption is not strictly accurate now, as symmetry does not preclude the
mixing of the fluorine $\pi'$-atomic orbitals (the $\pi'$-atomic orbitals have their
maximum amplitudes in the plane containing the four fluorine ligands) with
the fluorine hybrid atomic orbitals and the xenon $5s$ and $5p(\sigma\text{-})$ atomic
orbitals (see, for example, the table of symmetry orbitals as given by Malm
and co-workers[16]).

The tetrahedral configuration of fluorine ligands will give rise to a spatially

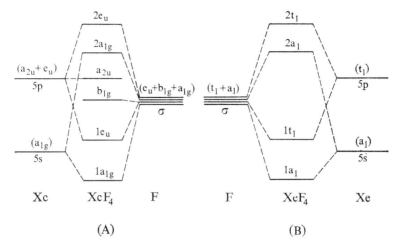

Fɪɢ. 4.12. Schematic molecular orbital energy level diagrams for $XeF_4$: (A) for the square planar arrangement, and (B) for the tetrahedral arrangement of ligands. Note that the atomic orbitals are again limited to the xenon $5s$ and $5p$ atomic orbitals, and four suitably disposed fluorine hybrid atomic orbitals.

degenerate ground state, and will therefore be unstable with respect to the vibrational motions of the nuclei. The square planar configuration, with the closed-shell electron configuration $1a_{1g}^2 e_u^4 b_{1g}^2 a_{2u}^2 2a_{1g}^2$, is clearly preferred. Some support for these conclusions is available from the results of the semi-empirical molecular orbital calculations of Lohr and Lipscomb.[17] They found that twice the sum of the occupied orbital energies for the tetrahedral and square planar configurations were $-37\cdot10$ a.u. and $-37\cdot26$ a.u., respectively, for an assumed Xe–F bond length of 0·24 nm; this favours the square planar configuration but, as mentioned previously, there are difficulties attending the use of sums of orbital energies as measures of total energies.

Although the ground state for square planar $XeF_4$ is expected to be spatially non-degenerate, there appears to be a possibility of vibronic inter-action involving the $^1E_u$ state, arising from the excitation $2e_u\leftarrow2a_{1g}$. However, the energy of this excited state will not be accurately given unless allowance is made for the involvement of xenon $5d$ atomic orbitals—or even $4f$ atomic orbitals. This follows from the observation that the $2a_{1g}$ and $2e_u$ energy levels are raised in energy as a result of the xenon–fluorine interactions, thereby making interaction with the xenon $5d(a_{1g})$ and $4f(e_u)$ atomic orbitals more significant. The inclusion of xenon $5d$ atomic orbitals in the LCAO expansion is expected to be more important, and this will lead to a widening of the energy gap between the $2e_u$ and $2a_{1g}$ energy levels; this in turn will lessen the significance of the effects of vibronic interaction. In this context, it is interesting that the semi-empirical molecular orbital calculations of

Yeranos,[18] based on the use of xenon $4d$, $5s$, $5p$, $5d$ and fluorine $2s$, $2p$ atomic orbitals, predicts that the highest energy-filled molecular oritbal ($a_{2u}$ symmetry) is separated by a large energy gap of 0·8 aJ from the unoccupied $e_u$ molecular orbital.

The inclusion of $5d$ orbitals also decreases the extent of formal charge transfer, by allowing the previously non-bonding molecular orbital, concentrated on the ligands, to encompass the xenon atom. These conclusions are similar to those found from the application of the valence bond model to $XeF_4$.

(c) *Xenon hexafluoride*, $XeF_6$. In the case of $XeF_6$, with $5d$ orbitals excluded, it is natural to assume an octahedral disposition of the fluorine ligands initially. The expected pattern of molecular orbital energy levels is shown in Fig. 4.13. The seventh pair of valence electrons is seen to occupy the

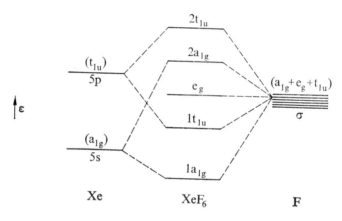

Fig. 4.13. A schematic molecular orbital energy level diagram for $XeF_6$, derived on the basis of contributions from xenon $5s$, $5p$ atomic orbitals and six fluorine $\sigma$-hybrid atomic orbitals.

$2a_{1g}$ (antibonding) molecular orbital, which will be concentrated mainly on the ligands. The twelve electrons occupying the $1a_{1g}$, $1t_{1u}$ and $e_g$ molecular orbitals give rise to the same situation that was found in SF—for it is the occupancy of these molecular orbitals which enables six localized one-electron functions to be formed, which, when doubly occupied, describe the six Xe—F bonds. So, just as with $SF_6$, the basic $e_g$ one-electron orbitals, which are required for describing the $\sigma$-bonding, are constructed out of ligand orbitals, and not excited orbitals of the central atom. Also, if the electrons in the $1a_{1g}$, $1t_{1u}$ and $2a_{1g}$ molecular orbitals are assumed to be shared equally between xenon and the ligands (it makes no difference if the $1a_{1g}$ and $2a_{1g}$ orbitals are concentrated preferentially on xenon and ligands, respectively), the molecular orbital model, in common with the simple valence bond model, requires a large formal transfer of charge from the central atom to the ligands.

It would appear from Fig. 4.13 that the electron configuration

$$1a_{1g}^2 1t_{1u}^6 e_g^4 2a_{1g}^2$$

would describe the ground state of the molecule. And, in view of the totally symmetric nature of this state, a regular octahedral structure for $XeF_6$ would be expected. But, unfortunately, just as in $XeF_4$, there will always be some uncertainty in this conclusion because, as seen earlier in Chapter 3 on transition metal ion complexes, it is very difficult to predict the electron configuration of lowest energy from the molecular orbital energy level diagram—particularly when the levels are not greatly separated in energy. A configuration interaction calculation is clearly required. In this respect, it should be noticed that the single electron excitation $t_{1u} \leftarrow a_g$ will give rise to low-lying singlet and triplet $T_{1u}$ molecular states. States with this spatial symmetry also arise from the single electron excitation $t_{1u} \leftarrow e_g$, and an additional $^1A_{1g}$ state arises from the double excitation $t_{1u} \leftarrow a_{1g}$; $t_{1u} \leftarrow a_{1g}$. Thus, in parallel with the discussion of the valence-bond model, it is found that even if the state arising from the electron configuration $1a_{1g}^2 1t_{1u}^6 e_g^4 2a_{1g}^2$ is the one of lowest energy (and this is not absolutely certain), there is still a group of low-lying excited electronic states, some of which are degenerate. And, as already seen in Chapter 1, these states can interact with the non-degenerate ground state through the excitation of vibrational motions, thereby causing a distortion of the ground-state ligand configuration. It is highly likely that a proper treatment of this dynamic model will require the inclusion of the effects of spin-orbit coupling, as these are expected to be significant for the xenon atom.

A more adequate description of the polarization of xenon by the ligands is obtained by extending the basis set of atomic orbitals to include $5d$ and perhaps $4f$, or even $6s$, atomic orbitals on xenon. The effect of including the xenon $5d$ and $4f$ atomic orbitals is shown schematically in Fig. 4.14. The $4f$ orbitals, in particular, appear to be important, because their presence leads to a decrease in the separation between the $2t_{1u}$ and $2a_{1g}$ molecular orbital energy levels. This should be contrasted with the situation in $XeF_4$ where there is also a possibility of a low-lying degenerate electronic state.

One further point ought to be mentioned about the molecular orbital energy level diagrams for $XeF_2$, $XeF_4$ and $XeF_6$; the patterns of energy levels have been estimated on the basis that the ionization potential of an electron in a fluorine $\sigma$-hybrid orbital is greater than that of an electron in a xenon $5p$ atomic orbital. It was also assumed that there was not too much charge transfer. However, it should now be apparent from both the molecular orbital and valence-bond treatments of the bonding, that there is a large drift of electron density from xenon towards the region of the ligands. This has the effect of preferentially lowering the energies of those molecular orbitals

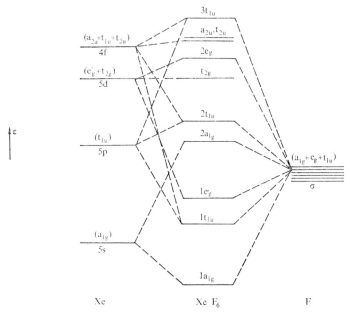

Fig. 4.14. A schematic molecular orbital energy level diagram for XeF₆, derived on the basis of contributions from xenon 5s, 5p, 5d and 4f atomic orbitals, and six fluorine σ-hybrid atomic orbitals.

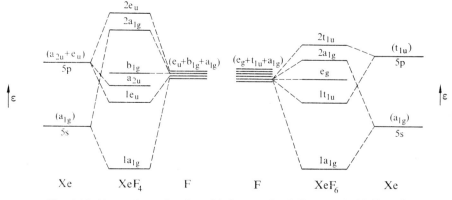

Fig. 4.15. Alternative molecular orbital energy level diagrams for XeF₄ and XeF₆ which may be more consistent with a significant amount of charge transfer from xenon to the ligands (as required by the simple model).

containing sizeable weightings of xenon atomic orbital coefficients. Thus, although it may be better, with hindsight, to construct the energy level diagrams for XeF₄ and XeF₆ in the manner shown in Fig. 4.15, nothing new is really gained as there is little difference between the new results and those given earlier. The main point in either approach is that it will be necessary to include the higher energy atomic orbitals on xenon in order to give an adequate description of the higher energy unoccupied molecular orbitals. For example, the $2t_{1u}$ molecular orbital in $XeF_6$ remains favourably placed for strong interaction with the higher energy xenon atomic orbitals, thereby decreasing the $2t_{1u}$–$2a_{1g}$ energy gap and increasing the relative proportion of xenon character in the molecular orbital. It is clear, therefore, that estimates of transition energies in $XeF_6$ will be seriously in error unless the contributions from the higher energy atomic orbitals on xenon are considered.

In conclusion, it remains to discuss why fluorine is so successful in stabilizing complexes of xenon. The reason appears to be the same as that found in the discussion of sulphur- and phosphorus-containing molecules; fluorine is a good polarizing ligand, which is also capable of accepting charge. The latter attribute also ensures that the $d$ (and $f$) atomic orbitals become more accessible; and both effects will tend to favour compact high energy orbitals on xenon. The presence of compact $d$-orbitals then facilitates the possibility of a back-transfer of charge from the ligands, so reducing the extent of charge transfer required initially. The fact that xenon has a lower ionization potential than either argon or krypton is clearly advantageous, in view of the extent of charge transfer which appears to be required to achieve an adequate bonding scheme. This receives some support from the fact that $KrF_2$ is the only fluorine compound of krypton which has been isolated (the higher fluorides requiring a greater extent of charge transfer). Finally, it should be emphasized again that the rare gas-containing molecules do not appear to be anomalous in any way—they fit into the series of molecules containing ten, twelve and fourteen valence electrons; the first two of which have been discussed at some length in Chapter 2. But, as seen in the discussion of $XeF_6$, the fourteen electron series is particularly interesting, because the description of the bonding requires a dynamic model for its complete elucidation.

## REFERENCES

1. PIMENTEL, G. C., *J. Chem. Phys.*, 1951, **19**, 446.
2. CLAASEN, H. H., SELIG, H. and MALM, J. G., *J. Amer. Chem. Soc.*, 1962, **84**, 3593.
3. SIEGEL, S. and GEBERT, E., *J. Amer. Chem. Soc.*, 1963, **85**, 240.
4. LEVY, H. A. and AGRON, P. A., *J. Amer. Chem. Soc.*, 1963, **85**, 241.
5. REICHMAN, S. and SCHREINER, F., *J. Chem. Phys.*, 1969, **51**, 2355.
6. HAMILTON, W. C. and IBERS, J. A., *Noble-Gas Compounds*, 1963, University of Chicago Press, Chicago, p. 195.
7. BURNS, J. H., AGRON, P. A. and LEVY, H. A., *Noble-Gas Compounds*, p. 211.
8. BOHM, R. K., KATADA, K., MARTINEZ, J. V. and BAUER, S. H., *Noble-Gas Compounds*, p. 238.
9. BARTELL, L. S. and GAVIN, R. M., *J. Chem. Phys.*, 1968, **48**, 2460, 2466.
10. DOW, J. D. and KNOX, R. S., *Phys. Rev.*, 1966, **152**, 50.
11. MITCHELL, K. A. R., *J. Chem. Soc.*, 1969, A, 1637.
12. SIDGWICK, N. V. and POWELL, H. M., *Proc. Roy. Soc.*, 1940, A **176**, 153.

13.  GILLESPIE, R. J. and NYHOLM, R. S., *Quart. Rev. Chem. Soc.*, 1957, **11**, 339.
14.  BARTELL, L. S., *J. Chem. Phys.*, 1967, **46**, 4530.
15.  BARTLETT, N., EINSTEIN, F., STEWART, D. F. and TROTTER, J., *J. Chem. Soc.*, 1967, A, 1190.
16.  MALM, J. G., SELIG, H., JORTNER, J. and RICE, S. A., *Chem. Rev.*, 1965, **65**, 199.
17.  LOHR, L. and LIPSCOMB, W. N., *J. Amer. Chem. Soc.*, 1963, 85, 240.
18.  YERANOS, W. A., *Mol. Phys.*, 1966, **11**, 85.

## BIBLIOGRAPHY

Reference 16: The Chemistry of Xenon.
*Noble-Gas Compounds*, 1963, University of Chicago Press, Chicago.
COULSON, C. A., *J. Chem. Soc.*, 1964, 1442: The nature of the bonding in xenon fluorides and related molecules.

# CALCULATION OF THE DIAGONAL MATRIX ELEMENT OF THE MOLECULAR ELECTRONIC HAMILTONIAN FOR A WAVE FUNCTION GIVEN AS AN ANTISYMMETRIZED PRODUCT OF ONE-ELECTRON FUNCTIONS (NOT NECESSARILY ORTHOGONAL)

Let the occupied spin orbitals be denoted by $\phi_i$ where, at this stage, it is unnecessary to specify which orbitals are associated with a particular spin function. The wave function is therefore represented by

$$\Psi = N \cdot \sum_{\hat{P}} (-1)^P \hat{P} \phi_1(1) \phi_2(2) \ldots \phi_n(n)$$

$$= N \cdot \begin{vmatrix} \phi_1(1) & \phi_1(2) \ldots \phi_1(n) \\ \phi_2(1) & \phi_2(2) & \cdot \\ \cdot & \cdot & \cdot \\ \cdot & \cdot & \cdot \\ \cdot & \cdot & \cdot \\ \phi_n(1) & \phi_n(2) \ldots \phi_n(n) \end{vmatrix} \qquad (AI.1)$$

where, as usual, $\hat{P}$ runs over all $n!$ permutations of the electron labels $1, 2, \ldots, n$; and $N$ is determined by the normalization condition

$$N \cdot \int \sum_{\hat{P}} (-1)^P \hat{P} \phi_1(1) \ldots \phi_n(n) \Psi \, d\tau = 1. \qquad (AI.2)$$

Now consider a typical permutation $\hat{P}$ in which the electrons labelled $p_1, p_2, \ldots, p_n$ occupy the orbitals $\phi_1, \phi_2, \ldots, \phi_n$. The integrations over the coordinates of the $n$ electrons are best performed by expanding $\Psi$ in (AI.2) in terms of the elements of column $p_1$ and their cofactors (see (AI.1)):

$$N^2 \cdot (-1)^P \int \phi_1(p_1) \phi_2(p_2) \ldots \phi_n(p_n) \sum_i A_{ip_1} \phi_i(p_1) \, d\tau$$

where $A_{ip_1}$ is the $n - 1 \times n - 1$ determinant formed by deleting the $i$th row and $p_1$th column from the determinant in (AI.1), and multiplying by $(-1)^{i+p_1}$. Integration over the coordinates of the electron labelled $p_1$ then yields

$$N^2.(-1)^P \int \phi_2(p_2) \ldots \phi_n(p_n) \sum_i A_{ip_1} S_{i1} \, \mathrm{d}\tau_{p_2} \, \mathrm{d}\tau_{p_3} \ldots \mathrm{d}\tau_{p_n}$$

$$= N^2.(-1)^P \int \phi_2(p_2) \ldots \phi_n(p_n) \times$$

$$\times
\begin{vmatrix}
\phi_1(1) \ldots \phi_1(p_1-1) & S_{11} & \phi_1(p_1+1) \ldots \phi_1(n) \\
\phi_2(1) \quad \phi_2(p_1-1) & S_{21} & \phi_2(p_1+1) \quad \phi_2(n) \\
\cdot \qquad\qquad \cdot & \cdot & \cdot \qquad\qquad \cdot \\
\cdot \qquad\qquad \cdot & \cdot & \cdot \qquad\qquad \cdot \\
\cdot \qquad\qquad \cdot & \cdot & \cdot \qquad\qquad \cdot \\
\phi_n(1) \quad \phi_n(p_1-1) & S_{n1} & \phi_n(p_1+1) \ldots \phi_n(n)
\end{vmatrix}$$

$$\mathrm{d}\tau_{p_2} \, \mathrm{d}\tau_{p_3} \ldots \mathrm{d}\tau_{p_n} \qquad\qquad \text{(AI.3)}$$

Expansion of the determinant in (AI.3) from column $p_2$, followed by integration over the coordinates of the electron labelled $p_2$, then gives

$$N^2.(-1)^P \int \phi_3(p_3) \ldots \phi_n(p_n) \times$$

$$\times
\begin{vmatrix}
\phi_1(1) \ldots \phi_1(p_2-1) & S_{12} & \phi_1(p_2+1) \ldots S_{11} & \phi_1(p_1+1) \ldots \\
\cdot \qquad\qquad \phi_2(p_2-1) & S_{22} & \qquad \cdot & \cdot \\
\cdot \qquad\qquad \cdot & \cdot & \cdot & \cdot \\
\cdot \qquad\qquad \cdot & \cdot & \cdot & \cdot \\
\cdot \qquad\qquad \cdot & \cdot & \cdot & \cdot \\
\phi_n(1) \quad \phi_n(p_2-1) & S_{n2} & \phi_n(p_2+1) \ldots S_{n1} & \phi_n(p_1+1) \ldots
\end{vmatrix}$$

$$\times \mathrm{d}\tau_{p_3} \ldots \mathrm{d}\tau_{p_n}$$

It is now easy to see that repeated application of this expansion and integration method then leads to the following contribution to the normalization integral

$$N^2.(-1)^P.(-1)^P
\begin{vmatrix}
S_{11} & S_{12} \ldots S_{1n} \\
S_{21} & \cdot \qquad \cdot \\
\cdot & \cdot \qquad \cdot \\
\cdot & \cdot \qquad \cdot \\
S_{n1} & \ldots\ldots\ldots S_{nn}
\end{vmatrix}
= N^2.\det |\mathbf{S}|.$$

As this result is independent of the choice of $\hat{P}$, it follows immediately that

$$\int \Psi\Psi \, \mathrm{d}\tau = N^2.\det|\mathbf{S}|.n!$$

yielding

$$N = \{n! \det |\mathbf{S}|\}^{-1/2}.$$

Now that $N$ has been defined, it is possible to calculate the energy integral

$$\int \Psi \hat{H} \Psi \, d\tau,$$

where, as already seen in Chapter 1, the molecular electronic Hamiltonian can be written in the form

$$\hat{H} = \sum_{i<j} \hat{h}_{ij} = \sum_{i<j} \left[ \frac{\hat{f}_i + \hat{f}_j}{n-1} + \frac{1}{r_{ij}} \right].$$

Consider the contribution of $\hat{h}_{12}$ to the energy integral:

$$\int \Psi \hat{h}_{12} \Psi \, d\tau = N^2 \int \left( \sum_{\hat{P}} (-1)^P \hat{P} \phi_1(1) \ldots \phi_n(n) \right) \hat{h}_{12} \times$$

$$\times \begin{vmatrix} \phi_1(1) & \phi_1(2) \ldots \phi_1(n) \\ \phi_2(1) & & \cdot \\ & \cdot & \\ & \cdot & \cdot \\ & \cdot & \\ \phi_n(1) & \phi_n(2) \ldots \phi_n(n) \end{vmatrix} \, d\tau. \quad (AI.4)$$

Now consider a typical permutation operator $\hat{P}$ which leads to the occupancy of the orbitals $\phi_1, \phi_2, \ldots, \phi_n$ by the electrons labelled $p_1, p_2, \ldots, p_n$, in which $p_1 = 1$ and $p_2 = 2$. The expansion of (AI.4) in terms of second-order cofactors associated with columns $p_1$ and $p_2$ then yields

$$N^2 . (-1)^P \int \phi_1(1) \phi_2(2) \phi_3(p_3) \ldots \phi_n(p_n) \hat{h}_{12} {\sum_{i,j}}' A_{i1, j2} \phi_i(1) \, \phi_j(2) \, d\tau$$

$$= N^2 . {\sum_{i,j}}' D_{i1, j2} \int \phi_1(1) \phi_2(2) \hat{h}_{12} \phi_i(1) \phi_j(2) \, d\tau_1 \, d\tau_2,$$

where $D_{i1, j2}$ is the determinant obtained from $\det|\mathbf{S}|$ by deleting rows $i$ and $j$ and columns 1 and 2, and multiplying by $(-1)^{i+j-1}$. However, there are $(n-2)!$ permutations associated with $p_1 = 1$ and $p_2 = 2$; and, in addition, the electron labels 1 and 2 may be associated with any other one of the possible pairs of orbitals. Hence the total contribution from $\hat{h}_{12}$ is

$$N^2 . (n-2)! {\sum_{p,q}}' {\sum_{i,j}}' D_{ip, jq} \int \phi_p(1) \phi_q(2) \hat{h}_{12} \phi_i(1) \phi_j(2) \, d\tau_1 \, d\tau_2. \quad (AI.5)$$

Each of the remaining $\hat{h}_{ij}$ gives rise to the same expression (AI.5), except that the integration variables are named as $i$ and $j$ rather than 1 and 2—a change which cannot alter the value of the integrals. Thus,

$$\int \Psi \hat{H} \Psi \, d\tau = \frac{1}{\det |\mathbf{S}|} \frac{(n-2)!}{n!} \frac{n(n-1)}{2} \cdot \sum_{p,q}' \sum_{i,j}' D_{ip,jq} \int \phi_p(1)\phi_q(2)\hat{h}_{12} \times$$
$$\times \phi_i(1)\phi_j(2) \, d\tau_1 \, d\tau_2$$

$$= \sum_{p,q}' \sum_{i,j}' \frac{1}{2} \cdot \frac{D_{ip,jq}}{\det |\mathbf{S}|} \cdot \int \phi_p(1)\phi_q(2)\hat{h}_{12}\phi_i(1)\phi_j(2) \, d\tau_1 \, d\tau_2$$

$$= \frac{1}{2} \sum_{p,q}' \sum_{i,j}' \frac{D_{ip,jq}}{\det |\mathbf{S}|} \left[ \frac{S_{qj}f_{pi} + S_{pi}f_{qj}}{n-1} + [pq/ij] \right] \tag{AI.6}$$

where

$$f_{pi} = \int \phi_p(1)\hat{f}_1\phi_i(1) \, d\tau_1$$

and

$$[pq/ij] = \int \phi_p(1)\phi_q(2) \cdot \frac{1}{r_{12}} \cdot \phi_i(1)\phi_j(2) \, d\tau_1 \, d\tau_2$$

In simplifying (AI.6), advantage can be taken of the symmetry properties of the $D_{ip,jq}$:

$$D_{ip,jq} = D_{pi,qj}; \quad D_{ip,jq} = -D_{jp,iq} = -D_{iq,jp} = -D_{qi,pj}$$

(these relations are true only for a choice of real functions $\phi_n$). Hence,

$$\int \Psi \hat{H} \Psi \, d\tau = \sum_{p,i} \frac{f_{pi}D_{pi}}{\det |\mathbf{S}|} + \frac{1}{2} \sum_{p,q}' \sum_{i,j}' \frac{D_{ip,jq}}{\det |\mathbf{S}|} [pq/ij] \tag{AI.7}$$

remembering that (see P.-O. Löwdin, *Phys. Rev.*, (1955) **97**, 1474, and references therein)

$$D_{ip} = \sum_{\substack{q \\ (i \neq j; \, p \neq q)}} D_{ip,jq}S_{jq} = D_{pi}$$

where $D_{pq}$ is the determinant obtained from $\det |\mathbf{S}|$ by deleting column $q$ and row $p$, and multiplying by $(-1)^{p+q}$. But from the properties of determinants (see Löwdin, op. cit.):

$$D_{ip,jq} = \begin{vmatrix} D_{ip} & D_{iq} \\ D_{jp} & D_{jq} \end{vmatrix} \times 1/\{\det |\mathbf{S}|\}$$

and so

$$\int \Psi \hat{H} \Psi \, d\tau = \sum_{p,i} \frac{f_{pi}D_{pi}}{\det |\mathbf{S}|} + \frac{1}{2} \sum_{p,q}' \sum_{i,j}' \frac{\{D_{ip}D_{jq} - D_{iq}D_{jp}\}[pq/ij]}{\{\det |\mathbf{S}|\}^2}$$

$$= \sum_{p,i} \frac{f_{pi}D_{pi}}{\det |\mathbf{S}|} + \frac{1}{2} \sum_{p,q}' \sum_{i,j}' \frac{D_{ip}D_{jq}}{\{\det |\mathbf{S}|\}^2} \{[pq/ij] - [qp/ij]\}$$

$$= \sum_{p,i} (\mathbf{S}^{-1})_{pi} \left\{ f_{pi} + \frac{1}{2} \sum_{\substack{q \\ (q \neq p)}} \sum_{\substack{j \\ (j \neq i)}} (\mathbf{S}^{-1})_{qj}\{[pq/ij] - [qp/ij]\} \right\}$$

where $(\mathbf{S}^{-1})_{ij} = D_{ji}/\det |\mathbf{S}|$.

However, the restrictions on the summations over $j$ and $q$ can be lifted, because the difference in two-electron integrals vanishes when either $j = i$ or $q = p$. The diagonal matrix elements of the molecular electronic Hamiltonian can therefore be written in the form

$$\sum_{p,i} (\mathbf{S}^{-1})_{pi} \left\{ f_{pi} + \frac{1}{2} \sum_{q,j} (\mathbf{S}^{-1})_{qj} \{ [pq/ij] - [qp/ij] \} \right\} = \text{tr} \ (\mathbf{fS}^{-1}) + \frac{1}{2} \text{tr} \ (\mathbf{GS}^{-1})$$

where

$$G_{ip} = \sum_{j,q} (\mathbf{S}^{-1})_{qj} \{ [pq/ij] - [qp/ij] \}.$$

Two special cases are of particular interest for a closed shell molecule: (i) the spatial orbitals for $\alpha$ and $\beta$ spins are different (this situation arises repeatedly in a valence bond calculation; but here, off-diagonal matrix elements of $\hat{H}$ must also be considered); and (ii) the spatial orbitals are orthogonal and doubly occupied with electrons having opposed spins ($\Psi$ now represents the simple molecular orbital wave function).

Consider case (i) first. If $\Psi$ is written in the form

$$\Psi = N \sum_{\hat{P}} (- 1)^P \hat{P} \phi_1(1) \phi_2(2) \ldots \phi_m(m) \bar{\psi}_1(m+1) \ldots \bar{\psi}_m(2m)$$

then $f_{pi} = 0$ unless both $p$ and $i$ label orbitals associated with the same spin function. Similarly, $[pq/ij] = 0$ unless both $p$, $i$ and $q$, $j$ label pairs of orbitals associated with the same spin functions.

In case (ii), the wave function is written as

$$\Psi = N \sum_{\hat{P}} (- 1)^P \hat{P} \psi_1(1) \bar{\psi}_1(2) \ldots \psi_m(2m-1) \bar{\psi}_m(2m)$$

$$= N \sum_{\hat{P}} (- 1)^P \hat{P} \psi_1(1) \psi_2(2) \ldots \psi_m(m) \bar{\psi}_1(m+1) \ldots \bar{\psi}_m(2m)$$

where

$$\int \psi_j \psi_i \, d\mathbf{r}_1 = \delta_{ji}$$

and so the energy expression takes the form

$$E = \frac{\int \Psi \hat{H} \Psi \, d\tau}{\int \Psi \Psi \, d\tau} = 2 \sum_{i,p} \delta_{ip} \left\{ f_{pi} + \frac{1}{2} \sum_{j,q} \delta_{jq} \{ 2(pq/ij) - (qp/ij) \} \right\}$$

$$= 2 \sum_i f_{ii} + \sum_i \sum_j [2(ij/ij) - (ji/ij)]$$

where

$$(pq/ij) = \int \psi_p(1) \psi_q(2) \frac{1}{r_{12}} \psi_i(1) \psi_j(2) \, d\mathbf{r}_1 \, d\mathbf{r}_2.$$

# CHARACTER TABLES FOR $O_h$ AND $T_d$ POINT GROUPS

| $O_h$ | $\hat{E}$ | $8\hat{C}_3$ | $3\hat{C}_2$ | $6\hat{C}_4$ | $6\hat{C}_2'$ | $\hat{J}$ | $8\hat{J}\hat{C}_3$ | $3\hat{J}\hat{C}_2$ | $6\hat{J}\hat{C}_4$ | $6\hat{J}\hat{C}_2'$ |
|---|---|---|---|---|---|---|---|---|---|---|
| $A_{1g}$ | 1 | 1 | 1 | 1 | 1 | 1 | 1 | 1 | 1 | 1 |
| $A_{2g}$ | 1 | 1 | 1 | −1 | −1 | 1 | 1 | 1 | −1 | −1 |
| $E_g$ | 2 | −1 | 2 | 0 | 0 | 2 | −1 | 2 | 0 | 0 |
| $T_{1g}$ | 3 | 0 | −1 | 1 | −1 | 3 | 0 | −1 | 1 | −1 |
| $T_{2g}$ | 3 | 0 | −1 | −1 | 1 | 3 | 0 | −1 | −1 | 1 |
| $A_{1u}$ | 1 | 1 | 1 | 1 | 1 | −1 | −1 | −1 | −1 | −1 |
| $A_{2u}$ | 1 | 1 | 1 | −1 | −1 | −1 | −1 | −1 | 1 | 1 |
| $E_u$ | 2 | −1 | 2 | 0 | 0 | −2 | 1 | −2 | 0 | 0 |
| $T_{1u}$ | 3 | 0 | −1 | 1 | −1 | −3 | 0 | 1 | −1 | 1 |
| $T_{2u}$ | 3 | 0 | −1 | −1 | 1 | −3 | 0 | 1 | 1 | −1 |

| $T_d$ | $\hat{E}$ | $8\hat{C}_3$ | $3\hat{C}_2$ | $6\hat{J}\hat{C}_2$ | $6\hat{J}\hat{C}_4$ |
|---|---|---|---|---|---|
| $A_1$ | 1 | 1 | 1 | 1 | 1 |
| $A_2$ | 1 | 1 | 1 | −1 | −1 |
| $E$ | 2 | −1 | 2 | 0 | 0 |
| $T_1$ | 3 | 0 | −1 | −1 | 1 |
| $T_2$ | 3 | 0 | −1 | 1 | −1 |

# SYMMETRY ADAPTED COMBINATIONS OF ATOMIC ORBITALS FOR AN OCTAHEDRAL MOLECULE (POINT GROUP $O_h$)

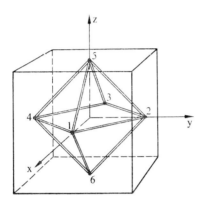

The $2s$, $2p_x$, $2p_y$ and $2p_z$ atomic orbitals on ligand $i$ are denoted by $s_i$, $x_i$, $y_i$, and $z_i$ respectively ($i = 1, 2, \ldots, 6$). The $4s$ and $4p$ atomic orbitals on the metal are denoted by $s_m$, $x_m$, $y_m$, and $z_m$, respectively.

| Free atom atomic orbitals | Symmetry adapted combinations of atomic orbitals (not normalized) | |
|---|---|---|
| **Metal** | | |
| $4p$ | $t_{1u}$ | $\begin{cases} x_m \\ y_m \\ z_m \end{cases}$ |
| $4s$ | $a_{1g}$ | $s_m$ |
| $3d$ | $t_{2g}$ | $\begin{cases} d_{xy} & (\zeta) \\ d_{xz} & (\eta) \\ d_{yz} & (\xi) \end{cases}$ |
| | $e_g$ | $\begin{cases} d_{z^2} & (\theta) \\ d_{x^2-y^2} & (\epsilon) \end{cases}$ |

| Free atom atomic orbitals | Symmetry adapted combinations of atomic orbitals (not normalized) |
|---|---|

**Ligands**

$2s$      $a_{1g}$   $s_1 + s_2 + s_3 + s_4 + s_5 + s_6$

$$t_{1u} \begin{cases} s_5 - s_6 \\ s_3 - s_1 \\ s_2 - s_4 \end{cases}$$

$$e_g \begin{cases} 2s_5 + 2s_6 - s_1 - s_2 - s_3 - s_4 & (\theta) \\ s_2 - s_3 + s_4 - s_1 & (\epsilon) \end{cases}$$

$2p(\sigma)$      $a_{1g}$   $-x_1 - y_2 + x_3 + y_4 - z_5 + z_6$

$$t_{1u} \begin{cases} -x_1 - x_3 \\ -y_2 - y_4 \\ -z_5 - z_6 \end{cases}$$

$$e_g \begin{cases} -x_1 + y_2 + x_3 - y_4 & (\epsilon) \\ 2z_6 - 2z_5 + x_1 + y_2 - x_3 - y_4 & (\theta) \end{cases}$$

$2p(\pi)$      $t_{1u} \begin{cases} x_2 + x_4 + x_5 + x_6 \\ y_1 + y_3 + y_5 + y_6 \\ z_1 + z_2 + z_3 + z_4 \end{cases}$

$$t_{2g} \begin{cases} z_1 - z_3 + x_5 - x_6 & (\eta) \\ z_2 - z_4 + y_5 - y_6 & (\xi) \\ x_2 + y_1 - x_4 - y_3 & (\zeta) \end{cases}$$

$$t_{2u} \begin{cases} -x_2 - x_4 + x_5 + x_6 \\ y_1 + y_3 - y_5 - y_6 \\ z_1 - z_2 + z_3 - z_4 \end{cases}$$

$$t_{1g} \begin{cases} -z_1 + z_3 + x_5 - x_6 \\ z_2 - z_4 - y_5 + y_6 \\ y_1 - x_2 - y_3 + x_4 \end{cases}$$

*Note.* The components of the $t_{2g}$ and $e_g$ symmetry adapted functions are also labelled, in brackets, according to the notation of Griffith (see ref. 3, Chapter 2).

# AUTHOR INDEX

AGRON, P. A. 155
ANDERSON, P. W. 135

BADER, R. F. W. 56, 60, 66
BALLHAUSEN, C. J. 97, 114, 118, 134, 135
BANDRAUK, A. D. 56, 60
BARTELL, L. S. 148, 155, 156
BARTLETT, N. 156
BASCH, H. 135
BAUER, S. H. 155
BEDON, H. D. 135
BERRY, R. S. 92, 97
BERTHIER, G. 65, 135
BERTONCINI, P. J. 66
BESAINOU, S. 65
BETHE, H. 98, 134
BISHOP, D. M. 64
BLINDER, S. F. 66
BOHM, R. K. 155
BONACCORSI, R. 65
BORN, M. 2, 5, 64
BOYS, S. F. 47, 65
BROWN, R. D. 86, 96, 97, 130, 131, 134
BUENKER, R. J. 65
BURNS, J. H. 155

CADE, P. E. 66
CAMPBELL, J. A. 97
CARLSON, K. D. 65
CHANDLER, G. S. 97
CHOPPIN, G. R. 135
CLAASEN, H. H. 155
CLEMENTI, E. 64
CONDON, E. U. 116, 134
COOLIDGE, A. S. 10, 64
COULSON, C. A. 35, 64, 78, 96, 97, 156
CRAIG, D. P. 77, 78, 79, 80, 96, 97
CROSS, P. C. 134
CRUICKSHANK, D. W. J. 79, 80, 96, 97

DAHL, J. P. 135
DAS, G. 65, 66
DAUDEL, R. 65
DECIUS, J. C. 134
DEL RE, G. 65
DIRAC, P. A. M. 35, 64
DOGGETT, G. 58, 65
DOW, J. D. 155
DUNN, T. M. 135

EINSTEIN, F. 156
ELLIS, D. E. 133, 134
ELLIS, M. M. 135

FENSKE, R. F. 132, 134
FEUERBAUCHER, D. G. 135
FIGGIS, B. N. 135
FRANCHINI, P. F. 30, 64
FREEMAN, A. J. 134, 135
FRIEDMAN, H. G. 135
FURLANI, C. 117, 134

GAVIN, R. M. 148, 155
GERBERT, E. 155
GIANTURCO, F. A. 78, 96
GILBERT, T. L. 66
GILLESPIE, R. J. 97, 156
GOLDBERG, H. 135
GONDAIRA, K. I. 135
GRAY, H. B. 135
GRIFFITH, J. S. 104, 119, 134, 149, 164

HAGSTROM, S. 64
HAMEED, S. 65
HAMILTON, W. C. 155
HALL, G. G. 13, 44, 64
HANDY, N. C. 47, 65
HARRIS, F. E. 64, 66
HARTMAN, H. 117, 134
HATFIELD, W. E. 65
HEITLER, W. 85
HELMHOLTZ, L. 43, 65
HENNEKER, W. H. 66
HERZBERG, G. 1, 65
HERZFELD, C. M. 135
HILLIER, I. H. 86, 87, 96
HINZE, J. 64, 77, 92, 96
HIRSCHFELDER, J. O. 65
HOPGOOD, F. R. A. 135
HORNER, S. M. 65, 135
HOWAT, G. 58
HUBBARD, J. 135
HUI, S. S. 65
HURLEY, A. C. 65
HURST, R. P. 30, 64, 79
HUTCHINGS, M. T. 135

IBERS, J. A. 155

JAFFÉ, H. H. 77, 92, 96
JAHN, H. A. 6, 64
JAMES, B. H. 131, 134
JAMES, H. M. 10, 64
JARRETT, H. S. 135
JOHANSEN, H. 97
JØRGENSEN, C. K. 65, 97, 125, 134, 135
JORTNER, J. 156
JUCYS, A. P. 65

KAMMSKAS, V. A. 65
KATADA, K. 155
KAVECKIS, V. J. 65
KAYAMA, K. 65
KIMBALL, G. E. 97
KIM, H. 64
KINCAID, J. F. 65
KING, H. F. 64
KLEINER, W. H. 125, 134
KLESSINGER, M. 29, 30, 64
KNOX, R. S. 155
KOIDE, S. 114, 134
KOŁOS, W. 1, 64
KOOPMANS, T. 47, 49
KOTANI, M. 65, 132, 134

LENNARD-JONES, J. E. 64, 65
LEVY, H. A. 155
LIEHR, A. D. 65, 110, 114, 118, 134, 135
LINDERBERG, J. 96
LIPSCOMB, W. N. 151, 156
LOHR, L. 151, 156
LONDON, F. 85
LONGUET-HIGGINS, H. C. 64
LOW, W. 118, 120, 134
LÖWDIN, P. O. 28, 30, 51, 58, 64, 65, 160
LUCKEN, E. A. C. 97

MACLAGAN, R. G. A. R. 97
MALM, J. G. 150, 155, 156
MARGENAU, H. 105, 134
MARSMANN, H. 97
MARTINEZ, J. V. 155
MATCHA, R. L. 65
McCLURE, D. S. 135
McKENDRICK, A. 65
McWEENY, R. 29, 50, 64, 65, 69, 72, 96, 111, 134, 135
MITCHELL, K. A. R. 79, 80, 96, 97, 139, 141–142, 155
MOFFITT, W. E. 65, 135
MOORE, C. E. 134
MULLIKEN, R. S. 51, 56, 57, 58, 59, 65
MURPHY, G. M. 105, 134
MUSHER, J. I. 65, 97

NESBET, R. K. 49, 65, 66
NEWMAN, D. J. 135
NEWTON, M. D. 51, 65
NYHOLM, R. S. 156

O'DWYER, M. F. 131, 134
OHNO, K. 65
ÖPIK, H. C. 64
OPPENHEIMER, J. R. 2, 5, 64
OSTLAND, N. S. 65
OWEN, J. 135

PARKS, J. M. 65
PARR, R. G. 64, 65
PAULING, L. 67, 96, 97
PEARSON, R. G. 135
PEEL, J. B. 86, 96, 97
PEKERIS, C. L. 1, 64
PETRONGOLO, C. 65
PEYERIMHOFF, S. D. 65, 129, 134
PHILLIPS, J. C. 125, 134
PIMENTEL, G. C. 136, 155
POPLE, J. A. 65
POWELL, H. M. 155
PROSSER, F. 64
PRYCE, M. H. L. 64, 114, 134

RADTKE, D. D. 132, 134
RANSIL, B. J. 65, 66
REICHMAN, S. 155
RICE, S. A. 156
RIMMER, D. E. 135
ROBERT, J. B. 97
ROBERTS, H. L. 97
ROBY, K. R. 131, 134
ROOS, B. 133, 134
ROOTHAAN, C. C. J. 49, 64, 66
ROS, P. 134
ROUX, M. 60, 65
RUMER, G. 85
RUNDLE, R. E. 82, 96

SACK, R. A. 64
SANTRY, D. P. 65, 86, 96
SCHONLAND, D. 97
SCHREINER, F. 155
SCHULMAN, J. M. 65
SCROCCO, E. 65
SEGAL, G. A. 65, 86, 96
SEITZ, F. 64
SELIG, H. 155, 156

SERBER, R. 64
SHERMAN, A. 64
SHORTLEY, G. H. 116, 134
SHULMAN, R. G. 134
SIDGWICK, N. V. 155
SIEGEL, S. 155
SINAI, J. J. 66
SLATER, J. C. 64, 65, 67, 96
SPINNLER, M. A. 79, 80, 96
STANTON, R. E. 64
STEWART, D. F. 156
STUART, J. D. 30, 64, 79
SUGANO, S. 125, 132, 134, 135

TAMRES, M. 97
TANABE, Y. 132, 134, 135
TELLER, E. 6, 64
THIRUNAMACHANDRAN, T. 77, 96, 97
THORNLEY, J. H. M. 135
THORSON, W. 135
TOMASI, J. 65
TROTTER, J. 156
TYREE, S. Y. 65, 135

VAN VLECK, J. H. 108, 134
VAN WAZER, J. R. 97
VEILLARD, A. 65
VERGANI, C. 30, 64
VISTE, A. 135

WAHL, A. C. 65, 66
WATSON, R. E. 135
WEBSTER, B. C. 79, 80, 96, 97
WHITMAN, D. R. 65
WILLETT, R. D. 90, 97
WILSON, E. B. 134
WOLFSBERG, M. 43, 65
WOLNIEWICZ, L. 1, 64
WYATT, R. E. 64

YERANOS, W. A. 135, 152, 156

ZARE, R. N. 100, 134
ZAULI, C. 78, 79, 80, 96

# SUBJECT INDEX

Adiabatic approximation  5
Al⁺, configurational mixing in  100

Al$^+$, configurational mixing in  100
Angular momentum operators  115–16
   spin  13, 14, 48
   step-down  116
Annihilation operator  48
Antisymmetrizing operator  12
Approximation(s)
   adiabatic  5
   Born–Oppenheimer  2–6
   integral  51
   LCAO  40
   perfect-pairing  23, 26, 74, 76, 77, 80
Atom(ic)
   charges  57, 58
   orbital(s)
     complete set of  18, 40
     $d$, size of  78, 92, 107, 139
     effect of symmetry operations on  22, 68–73, 102
     exponents  18
     extended basis set of  18
     $f$, symmetry transformation properties of  102
     Hartree–Fock  10, 40, 64, 78, 79, 138
     hybrid(s)  16, 22, 23, 27, 72–73 (see also Hybrids)
     non-orthogonality of  11, 19, 27, 28, 29, 79, 125, 133
     orthogonal  28, 30, 36, 43
     orthogonal transformations of  50
     $p$, $d$, symmetry transformation properties of  70–71
     phase convention for  76, 116
     population  57
     Schmidt orthogonalization of  29–30
     $\sigma$, $\pi$-type  14
     sizes of  74, 75, 77, 78
     Slater-type  10, 40, 61, 64, 78, 106, 108
     symmetrically orthogonalized  29, 43
     symmetry adapted combinations of  82, 127, 163–4
   units  xiii, 3

Born–Oppenheimer approximation  2–6
Bond
   order  59
   pair function  13, 27, 74
   population  57, 58
BH, electronic structure of  11–19

$BH_2$, electronic structure of  19–23

Character tables, for $O_h$, $T_d$  162
$CH_4$, electronic structure of  23–26
$ClF_3$, electronic structure of  92–96
CO  58
   electronic density functions for  55–56, 60–61
$Co(H_2O)_6^+$  130
$CoF_6^{3-}$  131
Cofactors  157, 159
Configuration(s)
   doubly excited  46, 54
   interaction (mixing)  45, 50, 77, 100, 118, 147
     for $MnO_4^-$  130; $XeF_6$ 153
   singly excited  45, 49, 130, 133
Correlation energy  47
Coulomb operator  37
Coupling, vibronic  108, 109–14, 118
$Cr^{3+}/Al_2O_3$  98
$Cr(H_2O)_6^{3+}$  125
$CrCl_6^{3-}$  132
Crystal field
   Hamiltonian  98, 100, 104–5
   theory
     appraisal of  122–6
     for $d^1$ complex ions  103–9; $d^2$ complex ions  114–21
$Cu(H_2O)_6^{2+}$  133
$Cu(NH_3)_6^{2+}$  133

Diagonal matrix elements of electronic Hamiltonian  157–61
Doubly excited configurations  46, 54
$d$ orbital(s)
   for Cl  96; P  92; S  74, 78–80, 85–88, 90; Xe  138–9, 143–9
   size of  78, 92, 107, 139

Effective Hamiltonian (operator)  4, 40, 45, 48, 49
Electron(ic)
   correlation  36, 46
   density function  52–64
   motion, separation of  2–6
   pair function  11, 12
   potential energy surface  5

169

Energy
　correlation　47
　electronic, invariance under orthogonal
　　transformations　39
　orbital　44, 79, 129, 130
　promotion　74, 75, 77, 80, 92, 96, 137,
　　145
　valence state　75–77, 86
Equivalent molecular orbitals　37, 84
Exchange operator　37

FCN　58
Fe(CN)$_6^{4-}$　130
FeF$_6^{4-}$　131
F$_2$　46
F$_2$O　63
Function(s)
　complete sets of　18, 40
　electron density　18, 52–64
　electron pair　11, 12, 13, 28, 53, 74
　　orthogonality of　31
　singlet spin　15, 21
　spin　11, 12, 20, 21
　　orthogonality of　21
　symmetry adapted　22, 24, 82, 111, 127,
　　163–4
　symmetrically orthogonalized　29, 43

Group generator　69

Hamiltonian (operator)
　crystal field　98, 100, 104–5
　diagonal matrix elements of　157–61
　effective　4, 40, 45, 48, 49
　model　7
　partitioning of　7
Hartree–Fock
　method(s)
　　extended　48
　　restricted　48–49
　　unrestricted　47, 133
　orbitals　10, 40, 64
　　for S　78, 79; Xe 138
HCN　58, 63
HCNO　63
HCCO$^-$, electronic excitation energies for
　129
He　1
Heitler–London method (structure, wave-
　function)　11, 29, 36
HF　58
H$_2$　1, 10, 11, 63
　electron density functions for　61–64

Hückel theory　132
Hybrid(s)
　for H$_2$O　27
　$sp$, for B　16, 22; S$^{++}$ 80–81
　$sp^3$, for methane　23
　$sp^3d^2$　72–73; for xenon 148
　$pd$, for xenon　137–8, 141, 146

Ionization potential　46
Integral approximation　51
　rotational invariance of　50–51

$j$–$j$ coupling　8–10
Jahn–Teller effect (coupling)　6, 108, 110

Koopmans' theorem　47, 49

Lagrangian multiplier　38
LCAO
　approximation　40
　molecular orbitals　42
LiH　30, 58, 79
Li$_2$　46
L–S coupling　8–10

Matrix
　representative　71, 72
　elements of Hamiltonian　19, 157–61
Methane　23–26
MnO$_4^-$　49, 129, 130
Model Hamiltonian　7
Molecular orbital(s)
　canonical　40
　energies　44, 87, 129, 130
　equivalent (localized)　37, 84
　Hartree–Fock　40
　Hückel　132
　LCAO　40, 42
　method (model)
　　for SF$_6$　82–88; transition metal ion
　　　complexes　126–34; xenon fluor-
　　　ides　149–55
　occupation number　55
　orthogonal transformations of　39, 41
　semi-empirical theory of　49–52, 132–3
　symmetry constraint on　48
　theory of
　　for closed shell molecules　36–47; for
　　　open shell molecules　47–49; for
　　　transition metal ion complexes
　　　126–34
　virtual　44, 45, 46, 83

Ne, Ne$^+$ 47
NaF 63
NH 58
NiF$_6^{4-}$ 126, 133
N$_2$, electron density functions for 55, 56, 60
N$_2$O 63
N$_3^-$ 63
Nuclear
  coordinates 3, 5, 6
  displacement vectors 111
  motion
    effective Hamiltonian for 4
    normal coordinates for 112
    wave functions for 2–6

Operator(s)
  angular momentum 13, 14, 48, 115–16
  annihilation 48
  antisymmetrizing 12
  Coulomb 37
  effective Hamiltonian 40, 48, 49
  exchange 37
  step-down 116
Orthogonal
  atomic orbitals 28, 30, 36, 43
  pair functions 31–36, 53–54
    electron density function for 53–54
  transformations on
    atomic orbitals 50
    molecular orbitals 39, 41

Pair function(s) 12, 28, 74
  bond 27
  orthogonal 31, 53
Pauling–Slater hybrids (method) 67, 72, 74, 84
Perfect pairing approximation 23, 26, 74, 76, 77, 80
PF$_5$, electronic structure of 92
Phase convention for atomic orbitals 76, 116
Promotion energy
  for B$^+$ 77; Cl$^+$ 96; P$^+$ 92; S 74, 75, 80, 86; S$^+$ 90; Xe 137, 145

Racah parameters 120
Renner effect 6
Representation
  reducible 23, 24, 48, 70, 71, 72, 83, 146
Russell–Saunders
  coupling 99
  term wave functions 8

SBr$_6$ 74, 79
Schmidt orthogonalization 29–30
Semi-empirical molecular orbital theory 49–52, 132–3
SF$_6$, SF$_5$, SF$_4$, electronic structures of 67–91
SH$_6$ 74
$\sigma$–$\pi$ separability 2
Slater-type orbitals 10, 40, 61, 64, 78, 106, 108
Spin
  angular momentum operators 13, 14, 48
  functions 11, 12, 20, 21
    orthogonalization of 21
    singlet 15, 21
  -orbit coupling 7, 8, 99
Step-down operator 116
Symmetry
  adapted functions 22, 24, 82, 111, 127, 163–4
  constraint, on molecular orbitals 48
  operation(s) 69
    on atomic orbitals 22, 68–73
    on hybrids 73
    for $T_d$, $O_h$ 69, 162
    matrix representatives for 71, 72
  orbitals 82, 127, 163
Symmetrically orthogonalized atomic orbitals 29, 43

Theorem
  Koopmans' 47, 49
  variation 3, 37, 38, 49, 67
  virial 18, 61
TiCl$_6^{3-}$ 132
TiF$_6^{3-}$ 103, 109
Ti(H$_2$O)$^{3+}$ 103, 108–9
Tm$^{2+}$/CaF$_2$ 102
Transition metal ion complexes, electronic structure of 98–135

Units, atomic xiii, 3

V$^{3+}$/Al$_2$O$_3$, electronic absorption spectrum of 118, 120
VCl$_6^{3-}$ 132
V(H$_2$O)$_6^{3+}$ 117, 126
Valence bond
  method 11–36; for BH 11–19; BH$_2^+$ 19–23; CH$_4$ 23–26; ClF$_3$ 92–96; H$_2$O 26–27; PF$_5$ 92; SF$_6$ 73–82; SF$_5$ 88–90; SF$_4$ 91; XeF$_6$ 144–9; XeF$_4$ 141–4; XeF$_2$13 7–41

Valence bond (*cont.*)
  structures
    covalent  18; for BH  11–17; CH$_4$
      23–26; SF$_6$  73; XeF$_6$  148; XeF$_4$
      141; XeF$_2$ 137, 140
    ionic  18; for BH  18; CH$_4$  24,  26;
      ClF$_3$  92–96;  H$_2$O  27;  PF$_5$  92–
      93;  SF$_6$  80–82,  85–86;  SF$_5$  91;
      SF$_4$  88–90;  XeF$_6$  144–7;  XeF$_4$
      142–4; XeF$_2$ 139–40
    long-bonded  17, 20, 140
    Pauling–Slater  68, 73–74, 137, 148

Variation theorem  3, 37, 38, 49, 67
Vibronic interaction (coupling)  108, 109–
  14, 148, 151
Virial theorem  18, 61
Virtual molecular orbitals  44, 45, 46, 83

Water, valence bond structures for  26–27
Wolsberg–Helmholtz method  49–50

Xenon fluorides, electronic structure of
  136–55